The Sex Life of Flowers

Bastiaan Meeuse and Sean Morris

The Sex Life of Flowers

Photographs by Oxford Scientific Films
Drawings by Michael Woods

Facts On File
NEW YORK

First published in the United States of America in 1984 by
Facts on File Publications
460 Park Avenue South
New York, New York 10016

Library of Congress Cataloging in Publication Data
Meeuse, Bastiaan.
    The sex life of flowers.
    Bibliography: p.
    Includes index.
    1. Pollination.   2. Plants, Sex in.   3. Flowers.
I. Morris, Sean.   II. Title.
QK926.M418   1984   582.13′04166   84–4044
ISBN 0–87196–907–6

This book was designed and produced by
The Rainbird Publishing Group Limited
40 Park Street, London W1Y 4DE

Text set by Oliver Burridge & Co. Ltd, Crawley, England
Color originated by David Brin Ltd, London, England
Printed and bound by Amilcare Pizzi s.p.a., Milan, Italy

Endpapers Flowers of the corn poppy (*Papaver rhoeas*)
open for one day only. They are self-sterile and rely on
insects to effect their cross-pollination.

Frontispiece The long-tubed flowers of red clover
(*Trifolium pratense*) offer nectar as a reward to their
bumblebee pollinators, but only those with a long
tongue, such as this *Bombus pennsylvanicum*, can
reach it.

Contents

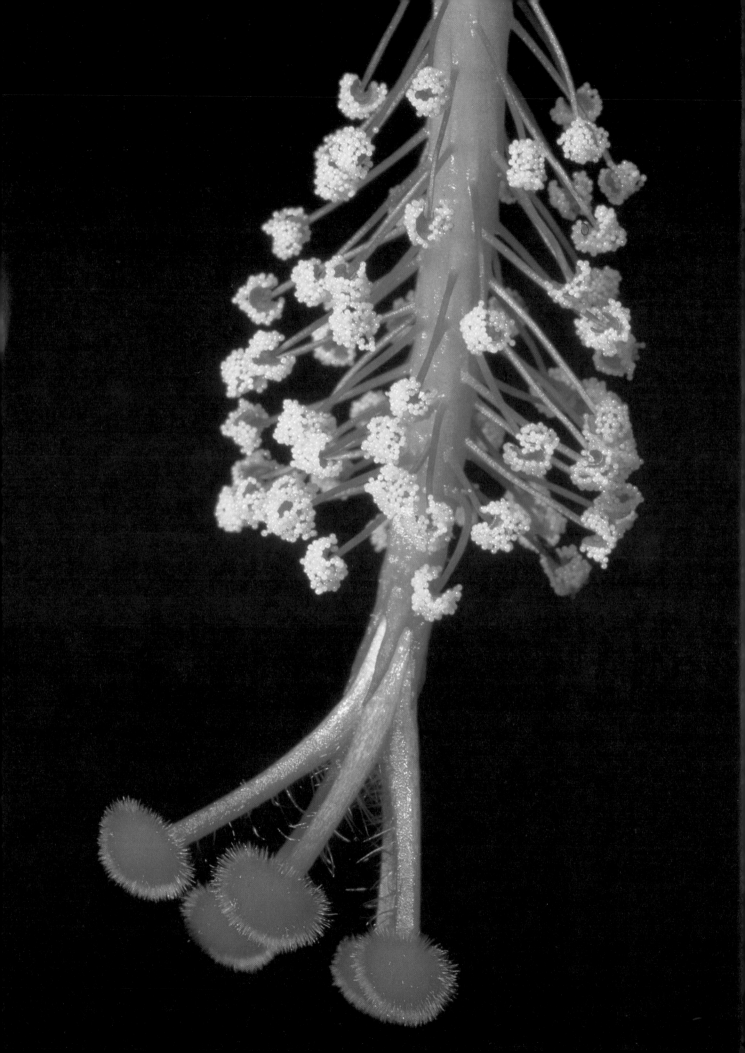

CHAPTER ONE

The evolution of plant sex

From time immemorial, flowers have appealed to our highest thoughts, ideals and emotions. We associate them with happiness, soft music and sunshine; we send them to our mothers on Mothering Sunday, and shower our lady singers with them after a successful aria. In dreaming up names for our flowers we have made a sincere attempt to reflect our innermost feelings: we call them 'forget-me-not', 'love-in-a-mist', 'baby's breath'. A little garden saxifrage is known in Holland as 'painter's sorrow', because it is simply impossible to capture its beauty on canvas.

It is a heart-warming thought that these good feelings go back to the very roots of our being. Not too long ago, the Neanderthal people were still pictured as hulking, low-browed brutes, not even capable of holding their bodies in an erect position. But a few years ago, the discovery of a Neanderthal burial site in a deep cave at Shanidar in Iraq altered this view quite drastically. Numerous pollen grains were found in the soil and these obviously could not have been brought in by the wind. Moreover, in a number of cases pollen grains of one type were found in small clusters – a clear indication that they had arrived in the form of intact flowers. Because the outer wall of each tiny grain is practically indestructible, pollen is easily recognizable even in soils that are millions of years old. Also the shape and size of the grains, and the beautiful fine sculpturing which they often possess, are different for different plant species. A pollen grain is to a botanist what a fingerprint is to a detective. By looking at a single grain under a microscope, an expert can often decide immediately what flower species it came from. It turned out that most of the pollen in the Shanidar cave came from showy, brightly coloured flowers – probably selected with great care on the surrounding plains. This suggests that the Neanderthal people recognized a spiritual aspect to life, and that their floral tributes to the dead were perhaps a means of expressing their feelings of reverence, respect and possibly even love.

Flowers have always been popular artists' models and have often been used in art to symbolize love, beauty and purity: the Virgin Mary holding an elegant Madonna lily is a well-known example. The discovery that flowers possess sex and are actually designed as organs for a plant's sexual reproduction was, therefore, greeted with some surprise!

The seed habit

The first man to recognize the role that flowers play in the sexual reproduction of plants was a seventeenth-century German professor, Rudolf Jakob Camerarius. He noticed that pollen is produced by parts of the flower

This showy array, presented by the coral hibiscus (*Hibiscus schizopetalus*) from tropical east Africa, is composed entirely of male and female sex organs. Pollen-laden stamens are united in a tubular column around a pendent, five-branched style terminating in sticky-haired pollen-receptors or stigmas.

now called the stamens and that, though sexless itself, it assumes a male role, while the slender pistil, which typically occupies the very centre of the flower, can be thought of as a female organ: in its more or less hollow bottom part it contains the ovules, the structures that will eventually give rise to the seeds.

The exact details of the act of sexual reproduction in plants were first recorded two centuries later by another German, Eduard Strasburger. Strasburger observed that when a grain of pollen finds itself on the broad tip, or stigma, of the pistil of an appropriate flower, it sends out a long tube, which grows down through the slender part of the pistil (the style) and delivers a nucleus, the sperm nucleus, to the egg cell present in the ovule, thereby fertilizing it (*see* p 23). The product of the fusion of sperm and egg cell develops into an embryo – the beginning of the next plant generation. It is only after fertilization that the ovule, held in the protective embrace of the mother plant, gradually transforms itself into a seed. This contains, within the seed-coat that represents the outer layer of the ovule, both the embryo itself and the endosperm, a special nutritious tissue for the future seedling to draw upon. In most cases, it is the seeds that act as the dispersal units of the flowering plant species, increasing the chance that seedlings will spring up far away from the mother plant.

It is fair to say that the phenomenal success of the flowering plants is due mainly to the 'invention' of seeds, and to the 'invention' of efficient methods for transporting pollen from the place where it originates to the appropriate pistils – the phenomenon we call pollination. Today, biologists recognize at least 300,000 species of flowering plant – 30,000 in the orchid family alone. They have colonized all sorts of environments, from deserts and arctic regions to windswept mountains, open plains and swamps. Some, such as the eelgrasses, have even gone back into the oceans, where plants originated hundreds of millions of years ago. Herein lies the justification for devoting this book to the sex life of plants.

In order to understand fully the process of sexual reproduction in flowering plants – the so-called 'higher' plants – we must start with these primitive ancestors and trace, as best we can, the evolution of the sexual habit from the earliest to the most recent plants. On the way, it will be important to remember that, like politics, biology is the art of the possible. Plants and animals sometimes have to forgo what may seem to us to be the most logical solutions to certain problems because they have to work with structures, inherited from their ancestors, that are not necessarily adequate for new jobs; they cannot start from scratch, and compromise becomes the order of the day.

In tracing the evolution of plant sex we are aided by the fact that a number of the most primitive types have survived, practically unchanged, right up to the present day. By studying these plants, and the far from negligible information that is available in the fossil record, we can piece together the details of a traumatic adventure, starting with single-celled, aquatic, 'primordial' plants, followed by more sophisticated, but still essentially primitive, aquatic forms, the algae. These gave rise to the first plants to invade the land, a step that necessitated not only the creation of new engineering solutions but also modifications of the methods of sexual reproduction. This stage is represented today by mosses and liverworts – land

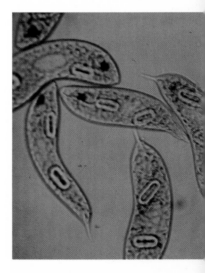

plants, but ones still able to exist only in moist places. Finally, as plants evolved towards greater stature and the ability to colonize a wider range of habitats, the flowering plants appeared. Tough, resilient plants, they were still dependent upon water, but they could live and reproduce on dry land using methods of sexual reproduction very different from those employed by their aquatic single-celled ancestors.

Before the invention of sex

A good starting point for our story of plant sex is, paradoxically, provided by a single-celled organism which, as far as we know, has no sex at all – the 'plantanimal' *Euglena* (from the Greek *eu* = good or genuine, and *glene* = eye). This microscopically small creature, which sometimes occurs in countless numbers in lakes and ponds, where it may cause 'water-blooms', is truly primitive. Since it possesses chloroplasts – small green chlorophyll-containing bodies that absorb and harness light-energy – it can build up its own carbohydrates by the process of photosynthesis. This makes it a plant. However, since it has a red 'eyespot' and can move around by thrashing a slender, whip-like appendage called a flagellum, it could also qualify as an animal.

In the centre of each *Euglena* cell we find the nucleus, the carrier of most of the cell's genetic information. The genetic material manifests itself in the form of a number of slender ribbon- or rod-like bodies called chromosomes, on which the genes – the factors that determine the genetic properties of the cell – are arranged in a linear fashion, like beads on a string.

There is only one way in which a population of *Euglena* can increase, and that is by each individual dividing lengthwise to form two new individuals. But before this event, the chromosomes in the nucleus are duplicated with uncanny accuracy. The nucleus then divides into two, an exact copy of the original linear arrangement of genes going into the nucleus of each of the two new *Euglena* individuals. The two 'daughter' cells are exact replicas of the mother cell. If you placed one *Euglena* in a pond and allowed it to multiply, you would probably end up, within a few weeks, with a population of countless millions of absolutely identical individuals. But not always. Occasionally, perhaps once in a million divisions, the gene-copying process goes awry; a mistake is made, and an individual is produced with a slightly different set of genes. A mutation has occurred. The new set of genes may make this individual better able to survive than its neighbours – it might, for instance, have a stronger flagellum, or a better-functioning eyespot, or a better mechanism for trapping light-energy. But beneficial mutations are rare. Many make no difference at all to the organism's viability: most prove fatal.

For *Euglena*, and in the final analysis for other organisms too, mutation is the only method by which new genetic forms can be produced. Mutations create variety in the offspring and the environment selects out the most viable types. But, as we have seen, mutations occur very infrequently, and this limits the rate at which a *Euglena* population can come up with a response to a change in the environment. Such a response is only possible when there is a sufficient number of superior types from which Nature can select. Once the favourable mutation is established in the new population,

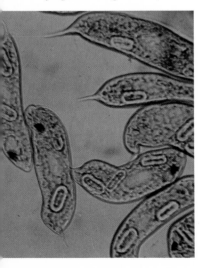

The microscopic single-celled organism *Euglena*, which creates the familiar 'green soup' in water-butts and ponds in summer, displays characteristics of both plant (the green discs being chloroplasts for fixation of light-energy, and the doughnut-like structures stores of carbohydrate) and animal (the whiplike flagellum facilitates active movement, and the orange-red, light-sensitive 'eyespot' is for guidance).

evolution will have occurred. But in our sexless *Euglena*, no convenient mechanism is available for spreading favourable mutations through a population, and for this reason it can evolve but slowly.

The meaning of sex

To understand how the 'invention' of sex influences the speed of evolution, let's look at another green unicellular creature, *Chlamydomonas*. Like *Euglena*, it can reproduce non-sexually. However, by means of a series of successive divisions, each *Chlamydomonas* cell can also produce eight, sixteen or even thirty-two small sex cells or gametes. These fuse in pairs, the members of each pair usually coming from different *Chlamydomonas* individuals. The fusion products are called zygotes, from the Greek word for yoke – something that combines, as in yoked oxen pulling a cart. A zygote has twice the number of chromosomes that a sex cell has; it is 'diploid' where the sex cells were 'haploid'. The zygote, capable of surviving adverse conditions, may lie dormant and unchanged for months, but will ultimately undergo a reduction-division to produce just four cells, each with only a single set of chromosomes.

These haploid cells will become new *Chlamydomonas* individuals. However, their genetic make-up will not necessarily be identical with that of the two gametes that fused to form the zygote. A reduction-division amounts to a reshuffling of the genetic cards in the deck. In addition, there is a good chance that the two members of each chromosome pair in the diploid cell will exchange sections of chromosome in a process called crossing-over. This too contributes to variability in the next generation. And genetic variation in a population is very desirable. When the environment deteriorates, a few individuals with superior qualities may still survive and leave offspring; and it is *their* type that will dominate future generations.

Generation cycles

Chlamydomonas illustrates the very important principle of alternation of generations. In its life-cycle, the emphasis is on the haploid phase; only the zygote is diploid. Sea lettuce (*Ulva*), a common seaweed often found washed ashore in the form of multicellular green sheets, has a 'fifty-fifty' cycle. In some of the sheets, all the cells are haploid. Together, they can be said to constitute the gametophyte, since the sheets are capable of forming gametes (which, of course, are haploid also). The zygotes produced when these sex cells fuse in pairs lead to diploid sheets through a series of ordinary divisions. These sheets represent the sporophyte generation. Sooner or later, reduction-divisions take place which lead back to haploid cells (spores), which can give rise to haploid sheets. In other algae, and also in land plants (with the exception of mosses), the diploid phase in the life-cycle is the dominant one. This makes excellent biological sense. In a haploid individual, a 'bad' gene resulting from a mutation will tend to have an immediate effect, and the individual will perish. In a diploid individual, the effect of a 'bad' gene in a particular chromosome may be masked or overruled by the presence of a good or advantageous gene in the chromosome's partner, that is in the other member of the chromosome pair. In this way the bad gene may be safely

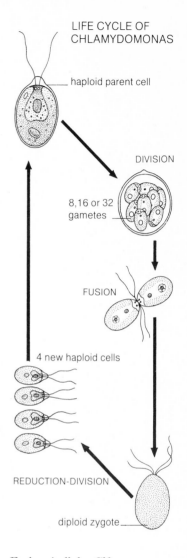

LIFE CYCLE OF
CHLAMYDOMONAS

haploid parent cell

DIVISION

8,16 or 32 gametes

FUSION

4 new haploid cells

REDUCTION-DIVISION

diploid zygote

Each unicellular *Chlamydomonas* undergoes a series of divisions, forming many small gametes which are morphologically identical to the parent organism. These later fuse in pairs to form diploid zygotes which may then lie dormant for some time before finally undergoing reduction-division that results in four new haploid organisms.

Anchored securely to rocks by a branched and suckered holdfast, kelps such as this *Laminaria digitata* make good use of the weight-reducing buoyancy of seawater by growing huge leaf-like organs for photosynthesis which are vastly larger than could be supported on land. Their motile reproductive spores disperse freely on the tide.

'stored' in the cell. In future generations, when the environmental conditions (and possibly other chromosomes) have changed, the 'bad' gene may be 'taken out' and used to advantage in the new situation.

The conquest of the land

As an environment, water has a lot to offer. Organisms living in it enjoy relative temperature stability and a reasonable exposure to light – at least in the upper layers – while at the same time they are protected from excessive radiation and desiccation. Being constantly bathed in a nutrient solution, the algae that live in streams and lakes, or in the ocean, have no need for roots to extract food from the soil. Extremely important also is the fact that in water weight hardly counts, because of the phenomenon of buoyancy. It is no coincidence that the largest creature our planet has ever seen, the blue whale, is an ocean-dweller. Its botanical counterparts, the giant brown *Macrocystis* seaweeds and the brown kelps (*Nereocystis*) of the American West Coast, have no trouble keeping their large leaf-like parts spread out for the efficient interception of the light rays that are essential for photosynthesis.

11

To leave the water's benevolent embrace, as the ancestors of our true terrestrial plants must have done, was therefore a hazardous and traumatic adventure. These ancestors must have been closely related to the green algae, as evidenced by the fact that these algae and the present-day terrestrials have the same photosynthetic pigments, the same food-reserve material (starch) and at least one cell-wall component (cellulose) in common. A simple comparison between a delicate green alga, totally dependent on water, and a modern tree, tall, erect and firmly rooted in the ground, suffices to show that the conquest of the land represents a true epic: breathtaking and so improbable that it seems to belong in the 'mission-impossible' category. Indeed, with the problems it posed in terms of temperature-fluctuation, desiccation, gravity, and the supply of water and nutrients, the terrestrial environment would seem to have been totally hostile to the ill-equipped settlers from the ocean.

The temptation to describe in detail the manner in which these problems were solved is great, but unfortunately we must resist it in a book such as ours, which has to concentrate on the sex life of plants. We know that the 'engineering problems' were solved, and that from that point on the early colonizers were in an excellent position to exploit the one factor that in the water-to-land transition had changed in a positive, beneficial way, and that was the availability of light. In water, light does not penetrate very far, and there can be no doubt that the lack of it quite often limits the rate of photosynthesis of algae. In terrestrial environments, on the other hand, light is often superabundant, and if other factors are adequate, large quantities of sugar and other carbohydrates can be produced through photosynthesis. Definitely, then, there is justification for the claim that many terrestrial plants are 'carbohydrate millionaires'. With sugar so cheap, most of the flowering plants can afford to use it as a bribe – in the form of nectar – for the insects and other animals that assist them in their sexual processes. This is all the more important because, in the final analysis, the conquest of the land, magnificent though it was, was still only a Pyrrhic victory. Stated baldly, the question is: how can two plant individuals of the same species, both anchored to the ground by their roots, interact sexually? It is here that the splendid principle of 'borrowed mobility' comes in. Bribed with nectar (and sometimes also with other rewards such as oil, perfume, pollen or food-bodies), insects and other animals act as go-betweens, carrying from one flower to another the precious pollen grains, which, as we have seen, can be regarded as the precursors of male sex cells. The *de facto* replacement of vulnerable sex cells by the much tougher pollen grains, combined with the use of carriers, is perhaps the neatest trick terrestrial plants have ever come up with. The bulk of our book will be concerned with it. However, we must first explain more precisely how the enormous gap between the algae and the flowering plants in the area of sex was gradually bridged in the process of evolution. Since we have to travel with seven-league boots, we will forfeit some detail and join the story part-way through – with the situation in ferns. (The more primitive bryophytes – the mosses, liverworts and hornworts – are not really relevant to the present discussion. They are still strongly dependent on water, do not contribute much to an understanding of flowers, and moreover are not the straight-line ancestors of ferns and flowering plants.)

Although able to withstand long periods of drought, ferns are entirely dependent on moisture for the germination of their spores. Woodland environments provide a host of damp niches ideal for the accomplishment of this phase of the fern's life-cycle.

The ferns

Thanks to the brilliant nineteenth-century discoveries of Hofmeister in Germany we know that in the life-cycle of the typical fern there is equal emphasis on the haploid and diploid phases. In some ways ferns are beautifully adapted to the dry conditions that so often characterize terrestrial environments; in fact, they take advantage of dry spells for the dispersal of their spores. Yet they are still totally dependent on liquid water for their sexual processes. It is because of this schizophrenic life-style that they are worthy of our close attention.

What most people think of as the fern plant is actually the diploid, spore-producing phase of the plant's life-cycle. The haploid spores are produced in structures called sporangia, which are combined in a number of small groups, each of which forms a brownish speck on the underside of a mature fern frond. As the spores are the dispersal-units in a fern's life-cycle, it is not surprising to find that the typical lollipop-shaded sporangium, as found in

13

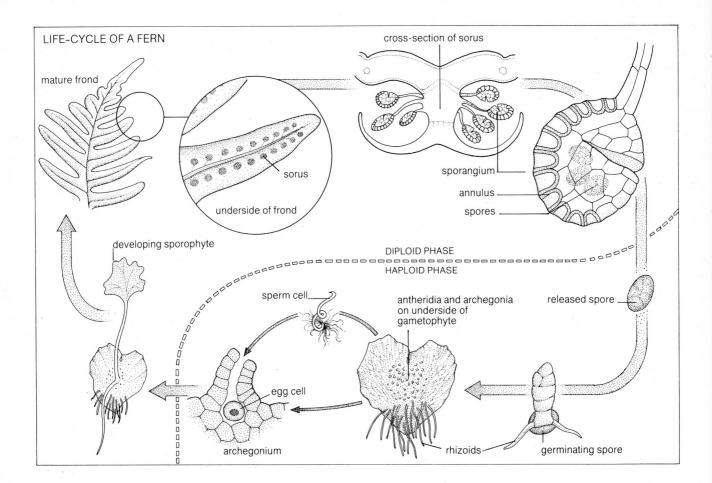

LIFE-CYCLE OF A FERN

mature frond

cross-section of sorus

sorus

underside of frond

sporangium

annulus

spores

developing sporophyte

DIPLOID PHASE

HAPLOID PHASE

sperm cell

antheridia and archegonia
on underside of
gametophyte

released spore

egg cell

archegonium

rhizoids

germinating spore

Polypodium species, has a highly specialized mechanism for their ejection. It is based on the presence of an annulus, a ring-shaped single row of dead, thick-walled, water-filled cells running around the periphery of the lollipop's head. Continuous water loss from the annulus cells as the result of dry weather conditions causes them to change their shape so that the ring straightens out and the central spore-mass breaks up into clumps. Ultimately, the water loss becomes so severe that the little bodies of fluid, which so far continued to fill each deformed annulus cell completely (because of the water's cohesiveness), suddenly break. The individual annulus cells, as well as the whole ring, snap back into their original shape and the spores are violently ejected to be further dispersed by the wind.

A fern spore must land in moist ground in order to germinate. There it develops into a tiny, flat, often heart-shaped structure, the gametophyte, composed entirely of haploid cells. Although only about a centimetre square, the fern gametophyte can lead a completely independent existence, for it is equipped with root-like organs called rhizoids, which take up water, and chloroplasts for photosynthesis. In all but a few genera, the gameto-phyte produces on its lower surface both male and female sex organs. By means of flagella the sperm cells from the male sex organs swim through the film of moisture that surrounds the gametophyte to fertilize the plant's egg cells *in situ*. The diploid zygote so formed germinates and eventually pro-duces a new spore-bearing fern plant, while the little heart-shaped gameto-phyte disintegrates.

Above Ferns are a familiar sight, but many people do not realize that they are seeing only half of the plants' life-cycle, the diploid sporophyte phase. Spore release initiates the onset of a second, less familiar, haploid gametophyte phase. Inspection of the underside of a mature frond (*below*) reveals rows of sori, which enclose the sporangia.

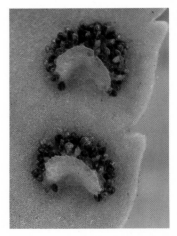

The flowering plants, or 'angiosperms'

The magnificent spore-dispersal mechanism of the ferns demonstrates that they have gone quite a way in developing a workable dry-land life-style. However, their dependence on liquid water for sexual reproduction has remained their Achilles heel. As we shall soon see, the flowering plants have managed to escape from this 'tyranny of water'. No longer do they employ free-swimming sperm cells. The manner in which they have done it represents a masterpiece of logistics and engineering, and it is altogether fitting and proper that in this book we make it the main course on the menu.

In the plant world, variety is the spice of life, and it is tempting to present our readers with representatives of the whole wide spectrum of different plant forms: horsetails, club mosses, water ferns, cycads, ginkgos, conifers such as pines, firs and cedars – even some fossil plants such as the seed ferns, which could teach us a great deal. However, we must cut through the underbrush and will restrict ourselves to the barest possible minimum. This is made possible because, almost unbelievably, a good grasp of the fern life-cycle alone already puts us in a position to understand, and interpret correctly, the situation in the flowering plants – the angiosperms. Firstly, though, we must take into account two new evolutionary wrinkles or twists that developed on the way to the flowering plants. The terminology involved appears daunting at first, but the process involved is essential to our understanding the importance of a pollen grain.

Certain ferns (or at least fern allies), rather than developing just one kind of sporangium and one type of spore, as *Polypodium* does, began to develop two different kinds: microsporangia and megasporangia. The microsporangia contain microspores that germinate into male gametophytes, that in turn produce male reproductive organs containing mobile sperm cells. The megasporangia, not surprisingly, contain megaspores which produce female gametophytes with female organs each containing one egg. Fertilization can occur only after the sperm cells have swum to an egg – there is still a water-dependent stage in the life-cycle.

The second twist is the elaboration of a principle we may call 'protective custody'. The liberated spores of a fern-like *Polypodium* have to face the cruel world with all its hazards. Would it not be a good idea *not* to liberate all the spores, but only the microspores, keeping the megaspores in the safe embrace of the plant that produced them? These megaspores could then each give rise to a megagametophyte and egg cells (or at least one egg cell) while maintaining their close association with the mother plant. After fertilization of the egg cell(s), even the young diploid embryo could enjoy the protection of the mother plant. The whole 'female' part of the life-cycle could thus be 'internalized'. Indeed, this is the situation we find in the modern flowering plant. The hollow bottom part of the pistil (which in the typical flowering plant occupies the centre of a flower) contains the ovules – each the equivalent of a megasporangium – surrounded by a jacket. In each ovule we find one egg cell: a situation analogous to the maternal care of the mammal embryo. It must also be obvious that a similar protective custody could not develop in the case of the microspores. They *must* leave the place where they are formed, otherwise it would be impossible for the sperm cells to find the egg cells, hidden in the ovules inside the pistil.

15

However, the claim can be made that even the microspores are subjected to some form of protective custody. Immediately after its formation, a microspore represents a single cell which is protected by a thin but strong wall. Within that protective shell the microspore nucleus can safely divide to produce what is essentially a two-celled microgametophyte. In the flowering plants this is usually the stage a microspore is in at the time it is liberated as a pollen grain. In a number of species the development is even further refined, and the liberated pollen grain may already contain three cells, or at least three nuclei.

With our understanding of ferns behind us, we can now reaffirm the statement made earlier that, strictly speaking, pollen grains have no sex. However, for all practical purposes they have taken over the role played by male sex cells in other organisms. We can now also see that a pollen tube with its contents, although not able to lead an independent life, can be legitimately compared with the microgametophyte of a fern.

The gymnosperms: 'plants with naked seeds'

Although the gymnosperms (ginkgos, cycads and conifers) should never be thought of as being on a direct line leading from ferns to flowering plants, they can help us greatly in understanding the latter, the angiosperms or 'plants with enclosed seeds'.

The gymnosperms best known to the general public are the conifers, which still dominate large areas of the northern temperate zone. They include trees such as pine, fir, hemlock, red cedar, spruce and redwood, and collectively are of enormous economic importance. As in the ferns, the diploid sporophyte is the most conspicuous generation – it forms the tree itself. The micro- and megaspores are produced in separate male and female cones. The microspores each develop into a minute four-celled gametophyte, a pollen grain, still enclosed within the original microspore wall. The female cones are much larger, and also more complex. Each of the thick cone scales bears on its upper surface two ovules. Although the ovule is protected by the tissues of the cone, the cover is far from complete; at the free end of the ovule we find a weak exposed spot called the micropyle, which potentially gives access to it. In spring, as the female cones are developing, the micropyle of each ovule sports a drop of sticky fluid. Pollen grains, liberated from the male cones and carried by the wind, stick to the drop like flies to fly-paper. The drop gradually shrinks as it dries out, sucking at least one pollen grain into the micropylar cavity and bringing it closer to the centre of the ovule. Here it is held for up to fifteen months, while the megaspores within the ovules gradually develop into female gametophytes; each of these finally produces egg cells, one of which will be fertilized by a sperm cell produced by the pollen grain. The sperm cells are 'delivered to the door', in this case by a tubular outgrowth of the pollen grain, the pollen tube, which forces its way through the tissue of the ovule and finally reaches the egg cell. A year at least lies between the act of fertilization and the moment the female cone received its ration of pollen. Hardly a case of sexual alacrity!

In a posthumously published letter, Charles Darwin referred to the origins of the angiosperms as 'an abominable mystery'. In his day the fossil record was still very poor, and it stands to reason that in its absence it was almost

Pollen grains, carried by the wind from neighbouring male cones, shower down over this female pine cone (*Pinus* sp). A sticky fluid exuded from each ovule catches the grains.

impossible to explain how nature could get from such situations as we have just described, with widely separated structures for the production of micro- and megaspores, exposed ovules and wind-pollination, to the most common flowering-plant situation in which male and female structures are combined in a hermaphroditic flower, the ovules are locked up in pistils and pollination is often brought about by animals.

A great deal of nonsense has been (and still is) written about 'the primitive angiosperm flower' – supposedly magnolia-like – and the role allegedly played by beetles in establishing the earliest insect-flower relationships. Admittedly, many present-day beetles are pollen-eaters, and there may have been many in the past. But why should such animals have been attracted by remote ovules which have no pollen to offer? They may have been interested in ovules as food, but obviously this would have prevented the formation of seeds rather than promoting it. Indeed, the origin of angiosperm reproduction has been explained by some biologists as a defence against voracious, roughhousing beetles.

The female 'flowers' of the European larch (*Larix decidua*), a deciduous conifer, are bright red and almost resemble true angiosperm flowers, often being nicknamed larch roses. Once fertilized by sperm cells produced by pollen grains released from the less showy male cones (*top of picture*), these female 'flowers' will develop into oval, brown, scaly cones.

The available evidence indicates that insects other than beetles (perhaps primitive flies?) are more likely candidates. The nutritious pollination-drop of the ovule may have given them their first incentive, and then, when angiosperm development eliminated this resource, nectar may have taken over. The present-day Gnetales (*Ephedra*, *Welwitschia* and *Gnetum*) offer excellent clues. Although still counted as gymnosperms, these plants should actually be referred to as hemiangiosperms, since they display several angiosperm features, such as the possession of filamentous stamens. As recent research has shown, both male and female structures produce nectar in some of the *Ephedra* species. There is a definite tendency towards sex reversal, as well as a tendency for the male and female organs to be combined in one structure. This much-neglected group of plants thus offers an excellent model of what may have happened in the distant geological past. At the same time it is important to keep an open mind to the possibility that the forebears of some modern wind-pollinated unisexual angiosperms were themselves always wind-pollinated and unisexual. After all, the gymnosperms are just that. Wind pollination in angiosperms is by no means always a secondary development.

Flexibility – the key to success

In gymnosperms there is a conservative trend, a tendency to preserve genetic types that have proved their value; in angiosperms, new gene combinations can be produced and tried out much more easily. This evolutionary versatility may frequently have led, even at an early stage, to features with definite survival value and it must have played an important role in the rise of the angiosperms and the establishment of their present dominance.

It is now fairly easy for us to imagine the situation 125 million years ago, when the pre-angiosperms had evolved the basic flower theme. The ancestors of today's insects – primitive beetles, flies and so on – clambered and flew from 'flower' to 'flower', plant to plant, consuming pollen, nibbling at ovules, probably doing almost as much harm as good. A very different picture from the situation found in animal-pollinated flowers today. Today's flowers are, in general, complex, precisely engineered devices that efficiently utilize the services of a variety of animals – insects, birds, bats, mice, even monkeys – as transporters of pollen, while the animals themselves have evolved ever more effective means of travelling from flower to flower, and for extracting food from them. What is more, many species of flowering plants use wind as their pollen-transporting agent, just as conifers do. Once the flower theme was established, the potential for diversification was enormous. This, in a nutshell, is to be the substance of the rest of this book, the manifold and bizarre ways in which flowering plants achieve fertilization.

CHAPTER TWO

The flower

So far we have attempted to trace the evolutionary processes that led from the humble beginnings of plant life to the 'invention' of the typical hermaphroditic flower – an event that probably occurred in the early Cretaceous period. Beautifully preserved fossil flowers very similar to those of our modern saxifrages have recently been found in Sweden, in rocks more than 78 million years old.

Having left behind them the tyranny of the water, the plants were faced with a new problem. Certain agents had to be pressed into service to transport their pollen grains from one flower to the female organs of another so that cross-fertilization could take place. Almost certainly these agents were primitive flying insects – mainly beetles and flies. Thus the flowering plants assured themselves of the priceless asset of 'borrowed mobility'. Moreover, in the 125 million years that have elapsed since the flower appeared, a truly astounding evolutionary development has taken place in which an entire army of mercenaries of every conceivable shape and size has been brought into action. In addition to insects, these 'helpers' include birds, bats and many other mammals. Some 'pre-angiosperm' plants, which at the dawn of the Cretaceous period depended on the wind for the dispersal of their pollen, probably continued in the same fashion, but others – after having gone through a stage where animals were the pollen-dispersing agents – *began* to depend on the wind, and even on water. As far as the animal agents were concerned, their evolution was in many ways dictated by the need to become ever more efficient flower visitors, while the plants, in a magnificent process of co-evolution, developed flowers that could dispense and receive pollen with ever-increasing efficiency. As we shall see, these events have in some cases led to partnerships in which there is an absolute mutual dependence of animal and plant.

But before we become too engrossed in the fine detail of how modern plants make use of their sexual middlemen, we must pause a while and learn the vocabulary of floral biology, and understand the basic structures common to all flowers – the building blocks used to fashion the countless variations of the fundamental flower theme.

The flower: what is it, and how does it work?

Even a relatively simple flower such as a buttercup turns out to be very much more than just a combination of male and female reproductive parts. To interpret it correctly, it is a great help to examine the structures that make up a typical, though imaginary, flower.

Starting at the outside we find a roughly circular whorl of sepals, which are usually green and fairly small in relation to the size of the flower as a

The colours, shapes and often scents of flowers, such as these cultivated poppies (*Papaver* sp), enable their pollinators to pick them out from their surroundings.

21

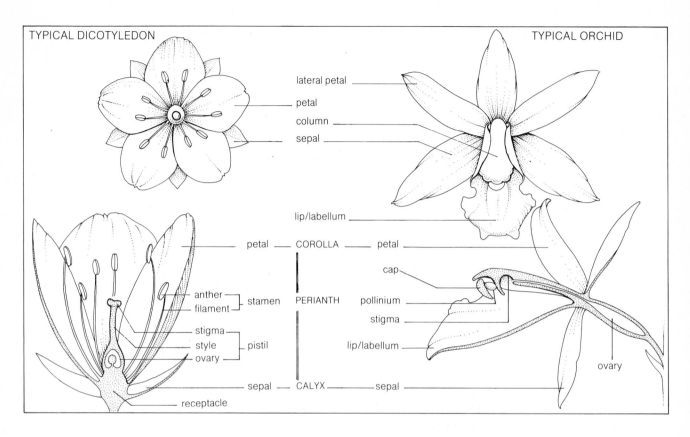

whole. Before the flower opens, it is the sepals, collectively called the calyx, that enclose and protect the developing bud. Inside the sepals are the petals, usually larger and more colourful. Together, the petals make up the corolla. In some flowers, for example in tulips, sepals and petals look very much alike, and since botanists are a playful lot at times, they call them tepals! Together, the calyx and corolla form the perianth – the colourful 'head' of the flower, whose main task, as we shall see, is to attract pollinators. Continuing in towards the centre of the flower, we encounter one or more whorls of stamens, which are the flower's male organs. Each stamen typically consists of a slender stalk or filament attached to the flower at its base and carrying on its free upper end a structure called an anther, which contains the pollen.

Finally, in the centre of the flower, are the female organs, or carpels. Individual carpels, as well as fused carpels which make up a pistil, consist of a basal ovary containing the ovules; a slender column-shaped structure called a style; and on the end of the style, the stigma, the function of which is to receive the pollen grains.

What happens when a pollen grain is deposited on a receptive stigma we already know: it sprouts a long, slender, tubular outgrowth that pierces the tissue of the stigma, worms its way down the length of the style and finally delivers one of its two sperm cells to an egg cell within the ovule. The latter is the structure that, after fertilization, gives rise to a seed.

All well and good. But in the real world of plants this logical, regular pattern of flower parts is not always so obvious. For instance, the function of coloured petals and sepals is often augmented by brightly coloured leaves or bracts. What is more, many plant species combine a number of individual

No two angiosperm flowers, whether monocotyledonous or dicotyledonous, have exactly the same structure nor, necessarily, the same components. Nevertheless, some broad generalizations can be made and the above diagrams show the structure of what may loosely be termed a 'typical' dicotyledonous flower and a more specialized monocotyledonous orchid.

flowers in what is called an inflorescence. Within such an inflorescence there is often a division of labour: the peripheral flowers may specialize as visual attractants and become sterile, while the central flowers (usually much less showy) serve for sexual reproduction. In overall appearance the inflorescence begins to look very like a normal 'simple' flower. This can be seen very clearly in some *Viburnum* species and in several members of the carrot family. The extreme case is found in the daisies, zinnias, dandelions, blue cornflowers, marigolds and other members of the aster family.

Of the 300,000 species of flowering plant, most depart in some way from our model of a typical flower. The number of sepals, petals, stamens and carpels, and the number of ovules per carpel, varies enormously from one species to another. For instance, while a poppy has only two sepals and four petals, a water lily may have many. The flowers of many orchids have only one stamen, while those of the baobab tree have as many as 2,000. Grasses usually have only one ovule per pistil, but in some orchid species the number is as high as a million. There may also be fusion of some of the flower elements, as we see in fuchsias and evening primroses with their well-developed floral tubes, or duplication of elements as we see in many members of the rose family. Changes in symmetry also occur, and many bee-pollinated species, for example, have rejected uniform radial symmetry for an elongated shape that also incorporates a landing area for the visitor. Some plants have even evolved separate male and female flowers, so reversing the very trend that initially (at least according to most floral botanists) brought about the evolution of the flower.

The biological function of most flowers is to secure the successful fertilization of their ovules by sperm cells derived from pollen from other flowers of the same species and to protect the developing ovules from attack by herbivores. Although many flowers are wind-pollinated, most botanists tend to think that this is a secondary feature and that the ancestors of today's flowering plants were all pollinated by animals. In this chapter, therefore, we will concentrate on those aspects of floral structure which are relevant to animal pollinators.

Attractants and rewards

Christian Konrad Sprengel (1750–1816), the father of floral biology, realized that bees and other flower visitors do not provide their pollen-carrying services free of charge. There must be a *quid pro quo*; that is, the plant must give them a reward in the form of something they need, either for their own survival or for that of their offspring. The most common reward is food of some sort, and this can be the pollen itself, a sugar-solution called nectar, or both. However, as we shall see later, there may also be others.

The adage 'It pays to advertise' holds good in the natural world, and many flowers announce their rewards by being conspicuous in colour, scent, size and shape, so making it easier for visitors to pick them out from their surroundings. (It would, of course, be a mistake to assume that animals perceive flowers exactly as we do: a honey-bee's senses of vision and smell, for example, are significantly different from ours.)

The size of flowers varies enormously, from the pinpoint flowers of duckweeds (*Lemna*) to the giant, one-metre-wide flowers of *Rafflesia*. As we might

POLLEN TUBE DEVELOPMENT

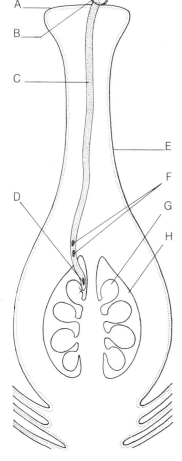

When a pollen grain reaches the stigma (A) it germinates (B) and sends out a pollen tube (C) which, controlled by the vegetative nucleus (D), grows down through the style (E). It finally reaches the ovary (H), where it delivers one of its gametes (F), the sperm nucleus, to the egg cell of an ovule (G).

expect, there is a positive correlation between the size of a flower and that of its pollinators. Flowers pollinated by birds or bats are definitely larger than those pollinated habitually by ants. However, the correlation is weak at best. The disproportion in size between bumblebees and the small forget-me-not flowers to which they are so partial borders on the farcical; the flower-stalks bend down almost to the ground under the weight of their insect visitors.

The shape, or rather the architecture, of a flower is largely determined by the number, size and design of its sepals, petals and stamens, and is dictated by a variety of needs. It is necessary to align the visitor so that it has the maximum chance of contacting the flower's sexual organs; to protect the anthers from damage by rain; to exclude visitors of the incorrect type or size; to protect the nutritious ovules against the onslaughts of herbivores; to provide various 'spring-loaded' pollen-depositing mechanisms; and, of course, to attract the visitors in the first place. To study the latter attribute of flower shape, divorced from the attraction exerted by colour and odour, is very difficult since the three typically work in concert, reinforcing each other. Experiments with honey-bees and bumblebees have shown that these animals prefer hollow flower models over flat or slightly convex ones. They are very poor at distinguishing between various two-dimensional shapes, but, like butterflies, they are good at perceiving differences in length of outline. In fact, honey-bees have an innate preference for figures with an elongated outline. Furthermore, when given the opportunity of visiting flower models shaped like leaves, they invariably select those with the more complex shapes, and will concentrate their attention on any teeth, serrations or lobes present.

Flower power: the controlled use of colour

To compare nature's way of doing things with the art of a composer or conductor is perhaps not totally ridiculous. In creating or re-creating a symphony, the human artist must rely on the perfect co-ordination and mutual co-operation of a large number of people and musical instruments; these form the elements, the units, which only through their combined action can produce the desired effect. What are the comparable 'units' in flowers?

First of all there are pigments, which can be present in dissolved form or as small solid particles. They can be offered side by side, in different cells, or they may be combined in a single cell. They can also be superimposed. Their effect can be modified by the structural peculiarities and the shape of the cells that contain them; by the presence of air-spaces in the petals; or by a waxy film or hairs on their surface. Furthermore, nature may also allow for a change in colour with time, or after pollination – something which in some cases can be done easily by a change in the acidity of the fluid in the cells.

Living plant cells, such as those found in flowers, can be thought of as tiny cellulose boxes with, on the inside, a thin layer of protoplasm that surrounds the large, central vacuole – a mass of water that holds in solution a whole array of chemical compounds such as sugars, plant acids, salts and often pigments. In most plants, the vacuolar pigments are anthocyanins and anthoxanthins: in the group of families collectively called the Centrospermae (but not in all of them) we find the so-called betacyanins, which

To its pollinators, and also to its human admirers, there can be no doubt that colour and scent are a flower's principal attractants. The operating colours of flowers are found mainly in the sepals and petals, and sometimes, as in the case of the bird-of-paradise flower (*Strelitzia reginae*), the colours of these two elements are wildly contrasting – orange sepals and blue petals – a seemingly inviting scheme in the eyes of the white-eye, here seen perched on the petals prior to feeding on the flower's nectar.

contain nitrogen in their molecules. The red pigment of beets, for example, and the purple one of *Bougainvillea* bracts, are both betacyanins.

Although the term anthocyanins suggests the colour blue (from its Greek derivation *anthos*, flower, and *kyanos*, blue), in fact these pigments range in colour from blue through purple to red, vermilion and scarlet, depending upon the particular anthocyanin we are dealing with, and also on the acidity of the cell sap. If the cell sap is acidic, chances are that the anthocyanin will appear red; if the sap is alkaline, it will be blue.

Another group of pigments, called the anthoxanthins, is responsible for colours ranging from pale ivory to deep yellow. Those that to humans appear ivory white absorb ultraviolet light very strongly and are therefore perceived as coloured (blue-green) by insects such as honey-bees, whose eyes are sensitive to ultraviolet light. When anthocyanins and anthoxanthins are present in the same petal, brown or reddish shades are produced.

Chlorophyll, the green pigment essential for the photosynthetic activity of leaves, is found in plant cells in small granules called chloroplasts. These also contain at least two additional pigments, carotene and xanthophyll, which are orange or yellow depending on their concentration. In the well-known South African bird-of-paradise flower (*Strelitzia*), carotenoid pigments are responsible for the intense colour of the orange parts. The same pigments are responsible for the striking orange colour of crocus stigmas; indeed those of *Crocus sativus* are dried and marketed commercially as the colouring and flavouring agent saffron.

White deserves special attention. Sometimes it represents no colour at all. In snow, for example, the white appearance is caused by the fact that light is scattered in all directions by the small air pockets it contains. Some flower petals too, such as those of daisies, contain air pockets in the intercellular spaces. When that air is eliminated with a vacuum pump, and replaced by water, the petals lose their whiteness and become colourless and transparent. 25

After 'white', 'black' has to be considered. Blackness indicates the complete absorption of all the visible-light wavelengths (colours). The almost cone-shaped epidermis cells that are found in the small black areas on the yellow petals of certain violets and pansies each contain two pigments: a yellow carotenoid is situated near the flat base, while a violet anthocyanin solution fills most of the remainder of the cell, towards the tip. Since the anthocyanin absorbs yellow and red, and the carotenoid absorbs blue, those regions appear very dark, almost black. The black patches at the base of the petals in red poppies and tulips demonstrate exactly the same principle, but here the two different pigments are not present in the same cells; instead, layers of blue-coloured cells underlie a surface layer of a deep purple tone.

Some flowers are able to produce the particularly warm colours and sheen of velvet because their epidermis cells, instead of forming a smooth, flat surface like a tiled floor, bulge out so that each one forms a steep little hill. They can legitimately be compared with the many short bundles of hairs which in velvet stick out at right-angles to the surface and which are responsible for the fabric's peculiar light- and colour-effects. This velvet effect is neatly demonstrated by the inside of the bract of a voodoo lily inflorescence, and also by gloxinias, pansies and snapdragons. The primary function of a velvety or 'papillate' epidermis is to serve as a light-trap for incident light, though in some flowers it plays a very different role. Certain terrestrial orchid species, for instance *Ophrys*, have flowers that are treated as sex partners by male wasps or bees, and the papillae may provide the insect with a tactile or visual stimulus. In other cases, for instance in poppies, the papillae may play a role in the rapid expansion of the petals that takes place when the flower bud unfolds.

The underlying reason for the seemingly endless variety of colours exhibited by flowers is based, in part, on the differences in colour vision of the various animal pollinators. Bees prefer blues and yellows, for example, while swallowtail butterflies prefer reds, and so on. But what of the intricate

Left The rich coloration of the petals of gloxinias (*Sinningia speciosa*), much prized by houseplant growers, is an excellent example of a 'velvet effect'. Innumerable tiny epidermis cells with modified surfaces (papillae) serve to trap incident light and create a colour effect which may attract a potential pollinator.

Different pollinators possess different colour vision – (*right above*) a pollen-laden buff-tailed bumblebee (*Bombus terrestris*) finds yellow appealing, for example. Similarly (*right below*), a swallowtail butterfly (*Papilio machaon*) responds to the orange-red colour of a marigold.

26

patterns of colour found in flowers? It could be that they simply make the flower more eye-catching and better able to draw the attention of the precious pollinator. Or could there be another function? We already know that pollinators visit flowers for purely selfish reasons, usually to help themselves to a high-calorie meal of nectar or pollen. Leaving aside for the moment the detailed discussion of the reward substances themselves, let us explore the idea that the patterns in and on the flowers are there to help the visitor find the reward more efficiently.

Prominent converging nectar guides on the petals of the grass-of-Parnassus (*Parnassia palustris*) serve to direct visiting flies to nectar glands at the flower's centre. Five three-pronged staminodes (sterile stamens), tipped with glistening, drop-like, false nectaries, also seem to lure flies.

Homing beacons and floral signposts

Science and religion are often given adversary roles, so there may be gentle irony in the thought that religion sometimes contributes to science. The life of Christian Konrad Sprengel certainly provides fuel for this idea. He was a deeply religious man, who saw the hand of the Good Lord in every little detail of nature. To him, even the most delicate hair on a flower, the smallest colour speck, had been put there by the Creator for a special purpose. It is not surprising, then, that it was he who discovered the role of nectar-guides, the signposts that help pollinating insects find the nectar of flowers – often deeply hidden within.

It all started with a humble little forget-me-not flower (*Myosotis palustris*). Sprengel's eye was quick to notice the yellow ring that surrounds the dark entrance to the flower and forms such a pretty contrast with the sky-blue of the petals. Could it be, he wondered, that the Good Lord, in His wisdom, had put it there to point the way to the sweet nectar inside? Earlier, while looking at the more exposed nectar of a wild geranium, he had already concluded that it is a treasure (and a reward for the bees) well worth protecting. The sweet drops at the base of the geranium petals are surrounded and partly covered by a ring of delicate hairs which shield them from dilution by rainwater and thus keep up their nutritional value. By looking at a large number of different flowers, Sprengel convinced himself that his ideas were correct. In many cases, he found small patches of contrasting colour on the petals near the flower's entrance, in others, patterns of dots or lines leading towards it. Collectively, Sprengel referred to them as *Saftmale*, from the German words *Saft* for sap or juice, and *malen*, to paint. In Anglo-Saxon countries they are called nectar-guides or honey-guides. Sprengel also noticed that flowers that are open at night, such as honeysuckle (*Lonicera*), certain silenes (*Silene*) and common evening primrose (*Oenothera biennis*), lack nectar-guides. This he considered understandable, for colour contrasts would not be of much use to pollinators that operate in semi-darkness.

Since Sprengel's day, an enormous amount of research has been done on the colour vision of pollinators – bees, flies, moths, butterflies, beetles, birds and mammals. While we will cover the conclusions of this research later we must mention here under the heading of flower patterns and nectar-guides that Sprengel's original hypothesis has been proved absolutely correct. From the pollinator's viewpoint many of the patterns displayed within flowers serve as strongly contrasting guidelines to direct them to the source of nectar, rather as bright yellow road-markings are used to direct traffic and patterns of lights lead an aircraft pilot to a safe touchdown. For instance, many bee-pollinated flowers contain (to bees) contrasting colour patterns of yellow and blue (Sprengel's forget-me-not); yellow and purple (eyebrights); orange and blue (certain lupins) and yellow and violet (pansies).

Nectar-guides are not restricted to colours that we humans can appreciate, but are often based on wavelengths of light that fall outside the range of human vision. If you set up a camera to simulate a bee's-eye view of the ultraviolet patterns in flowers, that is with a filter in front of the lens that allows only ultraviolet to pass to the film emulsion, the results are most startling, even though the photographs are in black and white. Meadow cranesbill (*Geranium pratense*), for instance, reflects ultraviolet from most of the petal surfaces, except for distinct nectar-guides radiating out from the flower centre. The flowers of evening primrose (*Oenothera biennis*) and marsh

To the human eye the flowers of this meadow cranesbill (*Geranium pratense*) seem quite uniform in colour with just faint streaks radiating from their centres (*below left*), but to an insect's eye, able to respond to ultraviolet wavelengths, distinct nectar guides are evident (*below right*).

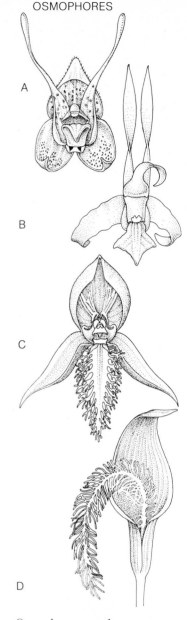

marigold (*Caltha palustris*) reflect ultraviolet strongly at the tips of the petals, but absorb it totally at their base, that is at the centre of the flower. To the human eye, the flower appears a uniform yellow colour.

Perfumes and scents

Smells are certainly an attribute of the environment that interests humans, and we know from personal observation that many animals have a keen sense of smell too. Dogs often seem largely preoccupied with scents; flies locate a putrefying carcass in no time, and wasps soon find fermenting, over-ripe fruit or an open pot of jam. Surely the pleasant (to us, that is) aromas given off by many flowers serve to attract their pollinators, just as their coloured petals do. Moreover, as aromas must be manufactured by the plant, and therefore form at least a modest drain on its energy resources, it is safe to assume that plants produce scents because they derive a specific advantage from so doing. Experiments have confirmed this conclusion. Many classes of pollinator have a keen sense of smell. Their preferences vary from pollinator to pollinator, as does their colour vision, and the scents given off by the flowers reflect these differences. As a result, we can often identify a flower by its scent alone. And so can many pollinators. Indeed, in some orchids that rely exclusively on one insect species for pollination, the scent is so diagnostic that one, and only one, species of insect will visit each species of orchid. Perhaps the most dramatic examples are found in the genus *Coryanthes*. There are about 30 species of *Coryanthes* in the Central American tropics, and it is thought that each one is pollinated by a different species of euglossine bee. The discrimination is based solely on scent. A one-to-one pollinator specificity is also found in several members of the genera *Stanhopea* and *Gongora* in Central America, and *Drakaea* in Australia. In most of these orchid species, artificial pollination of one species with pollen from another will produce perfectly viable hybrid offspring, but this rarely, if ever, occurs in nature.

Some pollinators can easily be taught to form an association between a given colour or smell and a reward such as honey or sugar-water. They will remember the connection quite well – in the case of the honey-bee for at least twelve days. However, while the bee's colour memory seems to be iron-clad and constant, the memory for odours varies with the time of day, following a distinct rhythm throughout the 24-hour cycle. This means that the insect remembers exactly the time of day at which it was taught to associate a particular odour and a food reward and will, over a period of several days, continue to look for the reward only at that particular time of day.

It goes without saying that this behaviour is very important in the everyday life of the pollinators that display it, because so many flower species open and close at a particular hour, or make their nectar and pollen available only during a short period of the day or night. Almost proverbial in this respect are the four-o'clock plant (*Mirabilis jalapa*), jack-go-to-bed-at-noon (a *Tragopogon* species that closes in the early afternoon) and evening primrose (*Oenothera biennis*), which opens in early evening. The beautiful light-blue flowers of chicory yield nectar only between 7 a.m. and noon. Wild mustard and certain dandelions offer their rewards around 9 a.m.;

Osmophores are odour-producing floral parts which are particularly common among members of the orchid and arum families. The slender, antenna-like palp osmophores of the orchid *Pleurothallis palpigera* (A) are modified inner petals. In *Dendrobium minax* (B) the inner petals are narrow and erect, providing a large surface area for odour evaporation, while in *Disa lugens* (C) evaporation is promoted still further by the provision of a fringed lip. The appendix of the arum *Arisaema fimbriatum* (D) develops the theme further by having paintbrush-like hairs, giving it an enormous surface area. (*Drawings A, B and C after S. Vogel, D after Engler, 1920*)

blue cornflowers around 11 a.m. Peak time for red clover, fireweed and marjoram is around 1 p.m., and for viper's bugloss and bachelor's buttons it comes at 3 p.m. The exact times will vary somewhat with the advance of the season, but honey-bees will always visit these flowers at exactly the right time.

Broadly speaking, a given flower species is characterized by its own special colour. It is even more sharply defined by its special odour, simply because there are many, many odour categories and relatively few colours. For this reason, odour may be even more important than colour in establishing a phenomenon known as flower constancy (which is very marked, especially in honey-bees and bumblebees). Flower constancy is a learned behaviour pattern which should never be confused with the inborn, specific pollination tie that may exist between a particular insect species and a particular species of flower, such as we have encountered in the case of euglossine bees and *Coryanthes* orchids. Flower constancy can be defined as the 'loyalty' which a pollinator displays towards the flowers of just one plant species. The pollinator is perfectly capable of pollinating other flower species, but will stick to its original behavioural pattern even when flowers that have more to offer in terms of nectar and pollen are becoming available in the landscape.

At first sight, such faithfulness may seem foolish. Yet it leads to a marvellous deal for both the pollinator and the plant species to which it is loyal. The plant benefits by not having its pollen wasted on flowers of another species, while the pollinator, having the opportunity to concentrate all its efforts on one type of flower, soon learns to handle it with maximum efficiency and speed. Other flower species that begin to appear may later be served by a slightly younger group of the same pollinator-species, namely by those individuals that emerge later and 'adopt' them.

It is easy to see that an insect such as a honey-bee or a bumblebee can actually develop constancy towards more than one plant species at the same time. If, for example, plant species A makes its nectar and pollen available between 9 and 10 a.m. and species B between 3 and 5 p.m., a bee can visit each at the proper time and exploit both species, while still keeping its book-keeping straight.

The time has now come to look at the other side of the arrangement – the rewards (or, to the cynic, bribes) that flowers offer their pollinators. Acceptable as 'legal tender' in the bargain between plant and animal are pollen, nectar, nutritious tissues, oils, waxes, resins and perfumes, with the understanding that a given pollinator is not interested in all of these, but concentrates on just one or two. Most butterflies, for example, are only interested in nectar. Furthermore, some flowers, especially in the fig family, offer their pollinators a favourable environment in which to raise their offspring, and a few orchids offer them shelter. Flowers of the orchid *Serapias*, found in damp open habitats in Israel, display a curious mimicry mechanism of pollination. While offering no rewards in the form of nectar or edible pollen, they offer bees (*Eucera*, *Andrena*, *Osmia* and *Tetralonia*) a kind of nest-replacement. Female bees of these solitary species sleep in holes in the ground; males may enter 'holes' mimicked by the orchid flower in search of the females, or merely to sleep. The 'hole' is in fact a short tube which has its entrance partially blocked by the column. Unsuccessful attempts by the

The flowers of the Mediterranean orchid *Serapias neglecta* offer nothing in the way of edible rewards to promote insect pollination, but instead provide certain solitary bees with a place to shelter for the night.

31

bee to get into the tube result in a pollinium being stuck to him. When later he tries to enter another 'hole', pollination is achieved. The bee is rewarded by finally forcing his way into a snug nest-hole for the night, and in the morning being warmed quickly, ready for his new day's work, as the heat of the rising sun penetrates the flower.

Pollen: nature's 'wholefood'

In our first chapter we tried to make it clear that in floral biology pollen plays a dual role: it is an indispensable link in the reproductive cycle of a plant, but in many cases we see that some of it is also sacrificed to benefit the pollinator. Pollen may well be the original bribe, for as we have seen earlier, pollen-eating beetles (and flies) probably stood at the cradle when the flowering plants appeared in the Cretaceous period. Pollen is a highly nutritious and well-balanced food material containing protein, sizeable amounts of starch, sugars, fat or oil, minerals, antioxidants and vitamins such as thiamin. It is also rich in free amino acids. There can be no doubt

Above A number of plants, such as *Papaver radicatum*, are heliotropic – their flowers 'suntrack' or rotate during the day so that they are always pointing towards the sun. This makes the flowers more attractive to their pollinators and raises the temperature of the flower's stigma, so speeding up the rate of germination of received pollen grains – advantageous in the Arctic, where ambient temperatures are low and the growing season short.

Right Flowers of the common mullein (*Verbascum thapsus*) each have five stamens – the lower two liberate fertile pollen; the upper ones provide insects with sterile food pollen.

therefore about its value as a food for insects and mammals. We shall see later that plant species offering pollen as a reward often produce it in huge quantities. The stamen count can be very high in the cactus, eucalyptus and mimosa families, and well-known pollen flowers from the temperate zone include the poppies *Papaver rhoeas* and *P. dubium*, wild roses (*Rosa* spp) and Scotch broom (*Cytisus scoparius*). Much pollen but very little nectar is also a feature of buttercups (*Ranunculus*) and many of the anemones (*Anemone* spp).

In order to minimize the loss of viable pollen to pollinators, some plants have found a nice compromise by producing, in one and the same flower, two types of anther; normal ones, which produce healthy pollen, and others that yield sterile or at least less viable but still very nutritious and probably tasty cells. While robbing these 'food anthers' of their contents, the insects bring about pollination, or at least they will be covered with viable pollen that can be used in another flower. Good examples of food stamens can be found in the genus *Cassia* in the pea family, in *Verbascum thapsus* (one of the mulleins), and in *Tibouchina* and other members of the Melastomataceae. In buffalo burr (*Solanum rostratum*) we again find two types of stamen, but most of the pollen from the food-anthers (which are 'milked' very efficiently by visiting bumblebees) is still viable. On the other hand, the feeding anthers of *Commelina sativa* do not contain any pollen but offer a milky fluid instead. Whether this is really consumed by visiting insects is still a matter for debate; some botanists see these anthers simply as advertising organs whose colour offers a striking contrast with that of the corolla; the normal anthers are rather dull in colour.

Some orchids have hit upon still another way to offer imitation pollen. In 1886, J. M. Janse found that the flowers of a tropical orchid (a *Maxillaria*) have on their lower lip a group of hairs which easily fall apart into individual cells, rich in starch. Similar 'food hairs' containing fat and protein were later found in the American orchid *Maxillaria rufescens*. Orchids cannot offer pollinators some of their pollen for food because, as we shall soon see, in this group of plants it comes in the form of neat little packages called pollinia, and any attempt to sacrifice a few grains would sacrifice the whole pollinium.

The efficiency with which certain pollinators collect pollen has long been a source of wonder, but recently the idea has been gaining ground that not all the credit for this should go to the animal pollinators. In the case of oil-seed rape (*Brassica napus*), for instance, it appears that pollen can actually jump across an air gap in order to attach itself to the body of a pollinator. Later it can, in the same fashion, jump from the animal on to an appropriate receptive stigma. It is quite possible that this remarkable mechanism, which employs static electricity, operates in other species as well. This, very simply, is how it works. It is now known that electrostatic charges can build up on the body of a foraging honey-bee, and that these charges may reach a magnitude of several hundred volts. The bee is therefore flying at the centre of its own highly charged electrostatic field. Most floral surfaces are well insulated, but the pistil is an exception: indeed there is a path of very low resistance leading from the stigma to the ground – almost like the earthing track of a lightning conductor. The result is that the bee's electrostatic field is attracted to the stigma – and with it the pollen. When a freshly killed pollen-laden honey-bee was electrically charged to +750 volts and then

33

moved towards a rape stigma impaled on an earthed pin, the pollen jumped from bee to stigma over a distance of more than 370 μm. (1 μm, known as a micrometre (formerly a micron) is one-thousandth of a millimetre.) When a similar 'bee electrode', without pollen, was moved towards an insulated anther, the pollen leapt from anther to bee over a distance of more than 600 μm. If these experimental results are also to be found in the natural world, they must greatly increase the chances of a bee picking up pollen – even if the animal fails to make contact with an anther. An operating range of half a millimetre is quite considerable!

Food-packages for pollinators

As an alternative to producing non-essential pollen, many plants produce solid food-bodies as a reward for their pollinators. Such food packages usually indicate that the pollinators are larger animals like beetles, birds or bats: nectar feeders, such as the majority of butterflies, would be unable to cope with them. Among the nicest examples are the cauliflower-like food-bodies at the tips of the pistils, stamens and staminodia (sterile stamens) of spicebush (*Calycanthus*), which is beetle-pollinated; the heat-producing ones attached to the carpels of the giant water lily (*Victoria amazonica*), also beetle-pollinated; and the big juicy bracts surrounding the inflorescences of the tropical liana *Freycinetia insignis*, on which the big bats known as flying foxes feast.

Nectar: the perfect fuel

It could be argued that when plants evolved flowers and embarked on their complex relationships with animals as their sexual middlemen, had nectar not been available as a potential reward-substance the plants would have invented it – or something very similar. It is, after all, an ideal substance for the purpose. It is easily manufactured; it can be produced in controlled quantities as demand requires; animal visitors can gather it simply by sucking up the fluid, which takes little time and effort; and as a solution of readily digested sugars it is a quickly assimilated source of energy.

Many scientists also believe that nectar was already being produced before flowers first appeared on the scene. Indeed the young leaves of bracken (*Pteridium aquilinum*) have distinct nectar glands (nectaries) even though they are not there for the benefit of pollinators. There is no reason to suppose that the ancestors of bracken (and of other modern ferns that have nectaries) lacked these organs. At the time when the flowering plants emerged from their fern-like ancestors, there may have been such extra-floral nectaries all over the place. It is quite conceivable that they were simply incorporated into the structures we call flowers at the time those structures developed.

One may, of course, ask why there was any need for nectaries of the *Pteridium* type to arise before there was a pollination function for them to fulfil. The Swiss botanist Frey-Wyssling has provided a physiological explanation. Young, vigorously growing leaves, he pointed out, have to be supplied with sugars and other building materials, and these are brought in by the phloem, the tissue which vascular plants generally use for transporting precisely such substances. When the young leaf begins to reach maturity

Above The flowers of South African proteas, such as *Protea magnifica*, are generally large and robust, being associated with fairly heavy pollinators. They produce copious quantities of nutritious nectar, and form a valuable food source for many species of sunbirds.

and its growth slows down, there may be a temporary excess of sugar, which the plant eliminates simply by excreting it. It cannot be a coincidence that the excreting organs, the nectaries, are practically always at the base of leaves or of organs derived from leaves. We find them in the flowering plants also, for example in flowering cherries, where they form neat, crater-like warts on the leaf-stalk just below the blade. Whatever their origin may have been, there can be no doubt that the appearance of a source of free sugar encouraged the development of relationships with insects, and later with other animals as well.

In the course of evolution these nectaries assumed an importance very much their own, with the result that in many modern flowers they appear as very complex structures. In some cases plants even use them as a means of discriminating against certain would-be visitors. The nectaries of love-in-a-mist (*Nigella*), for example, are so complicated that only exceptionally proficient insects such as honey-bees can get at the nectar within.

We have already seen how, during the course of evolution, plants 'discovered' that different colours, shapes and scents attracted different types

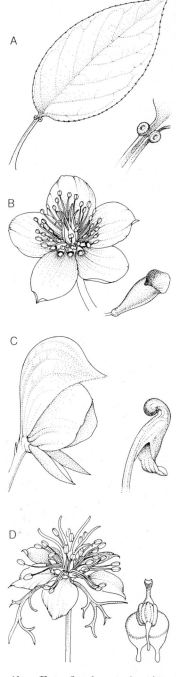

of pollinators. In much the same way, they also found that nectar itself could be modified – either in the quantity in which it was produced or in the proportions and types of sugars, amino acids, vitamins and minerals it contained. The result is that there are enormous differences in the nectars produced by modern flowers. Some of these adjustments are made to suit the specific needs of pollinators, but there are also a number of general considerations. For all nectar-producing plants there must be a balance between the quantity of nectar produced and the needs of the pollinator. If a given flower were to produce too copious a supply, the visiting animal might be satisfied at just one visit. It would not feel the urge to visit another flower for some time and consequently there would not be adequate cross-pollination. On the other hand, too small a nectar gift would not make it worthwhile for an animal to visit the flower at all. It is immediately obvious that large pollinators require larger nectar gifts than small ones, especially when they are warm-blooded creatures with a high rate of metabolism, such as hummingbirds and bats. There is also a connection between pollinator type and nectar concentration. Bird-pollinated flowers usually have a fairly dilute nectar, with a sugar concentration between 20 and 30 per cent. Birds simply cannot suck up a nectar that is too syrupy. On the other hand, horse chestnut flowers may have a sugar concentration as high as 70 per cent, with the result that their bee visitors have to dilute it with their own saliva before they can suck it up.

The observation that certain nectars – and consequently some honeys – can be poisonous has received a great deal of attention. Some of Xenophon's 10,000 soldiers, on their epic homeward trek from Persia, were incapacitated after consuming poisonous honey. In all likelihood, the culprit was *Rhododendron ponticum*. The nectar of California buckeye, a species of *Aesculus* and thus related to horse chestnut, is also often accused of leading to poisonous honey. However, in this case the real villain is the buckeye's pollen, some of which gets into the honey. Beekeepers are therefore well advised to remove their hives from any area where California buckeye is in bloom. In evaluating the situation, it is wise to remember that honey-bees are not native to California: the natural pollinators of buckeye were originally butterflies.

Nature's sweeteners

The main sugars found in nectar are glucose, fructose and sucrose. These three are usually found in various combinations, although there are some nectars which are totally devoid of sucrose, while others contain only this sugar. In terms of energy, it does not make much difference which sugars are present. The reason why it is often very difficult to be specific about the exact sugar combination of nectar is that it is very, very easy for the sucrose molecule to react with water and to fall apart into its two building blocks, glucose and fructose. This happens, for instance, when nectar is contaminated with certain yeasts, an event which is not at all uncommon in nature. There is a whole group of yeasts that thrive in nectar, and ants sometimes carry these from one flower to another, on their legs. At least some yeasts produce an enzyme, invertase or sucrase, which splits sucrose. Honey-bees also possess this enzyme, and consequently honey contains no sucrose. This helps in preventing, or slowing down, crystallization.

Above Extrafloral nectaries (A) are located on the leafstalks of cherries (*Prunus serrulatus*). Equally accessible to pollinators are the urn-like nectaries (B) of Christmas roses (*Helleborus* spp). Monkshoods (*Aconitum* spp) have concealed, horn-like nectaries (C), while love-in-a-mist (*Nigella*) has complex nectaries.

Right The bromeliad *Billbergia nutans* derives its common name Queen's-tears from the fact that thin nectar drips from floral glands, forming droplets close to the pollen-bearing anthers.

A very special relationship exists between the yellow loosestrife (*Lysimachia vulgaris*), a European bog plant, and bees of the species *Macropis europea* (*left*). While the loosestrife can survive in regions where the bee is absent, the reverse is not true. These bees are totally reliant on a particular oil for their grubs' nutritional requirements – this oil is supplied, as a pollination reward, only by the yellow loosestrife. The bee's feet have a velvety pad of dense hairs (*above*), which they use to mop up the oil from the flowers.

Oil-flowers

The falseness of the old adage that there is nothing new under the sun was beautifully demonstrated in 1969 when Stefan Vogel discovered that certain flowers offer fatty oils (glycerides) instead of sugary nectar to visiting bees. He listed five families of plants with at least some representatives that have oil-producing glands – or elaiophores, as he called them – in their flowers. That number has now risen to eight, with about 80 genera and approximately 2,300 species.

From sad experiences with wrecked oil tankers, humans have learned that oil is a nasty material to handle. However, the oil-collecting bees that have co-evolved with the oil-producing flowers are superbly equipped for the task. The American ones turned out to be closely related and comprise seven genera, of which we mention *Centris*, *Tetrapodia*, *Paratetrapodia* and *Monoeca*. In the Old World, only two genera of oil-collecting bees, *Macropis* and *Ctenoplectra*, have so far been found.

As might perhaps be expected, the oil-collecting apparatus in the American bees is very similar in the various genera and species. Special hairs and combs of hairs aid in the collection, transportation and temporary storage of the somewhat sticky oil. They are placed on the first, second and third pairs of legs, and although their number, size and exact position there varies, one can discern a common pattern. In some species, the hair-cushions on the front legs are so dense that they resemble sponges, and the insects do indeed use them as such, dipping them into the oil sometimes after real contortions to get into the right position. In other oil-bees, flattened and somewhat blade-like structures form the collecting apparatus. In general, it is only the females that possess the special hair-structures, and it is they that mix the collected oil with pollen (and perhaps in some cases with nectar also) to produce a most nutritious and protein-rich 'bee-bread' for their larvae.

Resins and waxes

Waterproof resinous substances can be very valuable as nest-building materials, especially in wet tropical countries. They are used by many bees and wasps, including honey-bees (*Apis*), stingless bees (*Meliponini*), *Centridini* (Anthophoridae), a number of leafcutters (Megachilidae) and several wasp genera. These animals collect the precious material from resin-producing plants, usually from the tips of broken twigs, leaf buds or wounds in the bark of trees. In view of this, it is surprising that nature has not utilized resinous materials more often as a reward in the pollination of flowers. So far, only four plant genera (*Clusia*, *Ornithidium*, *Eria* and *Dalechampia*) have been found to use this method, but only for *Dalechampia*, a large tropical genus which includes more than 100 species, has it been proved beyond doubt that the substance involved is actually a water-insoluble resin and not just a soluble thick gum. The showy flowering structure of *Dalechampia* is one of those inflorescences that look deceptively like a flower, and for that reason it is usually called a pseudo-flower. The real, small flowers are subtended by two large and often brightly coloured bracts which support a group of three sessile female flowers and another one of nine to twelve male flowers. The latter have their own little 'bractlets', but these are converted into secretory structures which together form one large and conspicuous gland, responsible (in most species) for producing the highly sticky resinous compound. Pollinator activity is limited by the fact that the large bracts remain closed around the flowers except for a short period of opening, which in the Mexican species *D. magnistipulata* lasts from 4.30 to 6.30 p.m. only.

Waxes are gathered by certain tropical bees for use as a vital nest-building material. Only a very few plant species are known to offer wax as a reward to their pollinators. Here a stingless *Trigona* bee visits the male flower of a dioecious Central American shrub, *Clusia uvitana*. The flower's golden wax secretions are clearly visible.

39

CHAPTER THREE

The incest taboo

In describing the methods by which flowers attract their pollinators and subsequently reward them for their services we have assumed all along that the pollinators' task is to transfer pollen from one plant to another – the very basis of sexual reproduction in higher plants. Scientists call it outbreeding. As we saw in Chapter One, outbreeding carries with it the likelihood of variation, a distinct advantage in a world of changing environments. Variation allows natural selection to mould the genetic basis of a species to cope better with the problems of survival and reproduction.

While outbreeding is so obviously an advantage, there are circumstances in which the opposite, self-fertilization or inbreeding, is a satisfactory or even superior method of sexual reproduction. We will cover these situations and their impact on the design and 'behaviour' of flowers in a later chapter. For the moment we will concentrate on the refinements that are found in many flowers to encourage outbreeding.

When the flower theme was invented by the forerunners of today's flowering plants and the male and female organs were for the first time brought together within one structure, a major problem was created. The animal visitors to those flowers were as likely to fertilize the female parts of that flower with pollen from the *same* flower as they were to bring in and deposit pollen from a flower visited previously. If a flower is fertilized by sperm from its own pollen, or indeed by that from pollen coming from another flower on the same plant, many of the advantages of sexual reproduction are lost, as the offspring exhibit very little variation. So we might expect plants to have evolved all kinds of mechanisms for promoting outbreeding, and indeed they have. They include self-incompatibility, that is chemical barriers that prevent pollen from germinating on the stigmas of the flower that produced it, or of other flowers from the same plant; spatial separation, or the setting-up of barriers between the male and female structures in one flower; separation in time between pollen release and female receptiveness in one and the same flower (dichogamy); separation of male and female flowers on one plant (monoecism); separation of the sexes by having the male and the female flowers on different plants (dioecism); and bisexual flowers in which anthers and stigmas are at different levels (heterostyly).

Spatial separation

As to spatial separation of male and female parts, excellent examples are found in the orchids and in the milkweeds (Asclepiadaceae). In both these groups, pollen is packed into fairly hard masses with a wax-like appearance, called pollinia. In the milkweeds these are stored away safely in the central

Self-pollination has a number of disadvantages with respect to the success of a given species, and for this reason various mechanisms have evolved to prevent this occurring. In the flowers of the rose-bay willow herb (*Epilobium angustifolium*) the lower, older flowers contain receptive stigmas, while the upper ones bear active stamens.

41

Left Pollinia, masses of pollen bound together by minute elastic threads and with an adhesive pad, are transported from one orchid flower to another by insects. They may adhere by a variety of mechanisms to several parts of the insect's body according to the particular species involved. In this case orchid pollinia have become glued to the coiled proboscis of a hawkmoth.

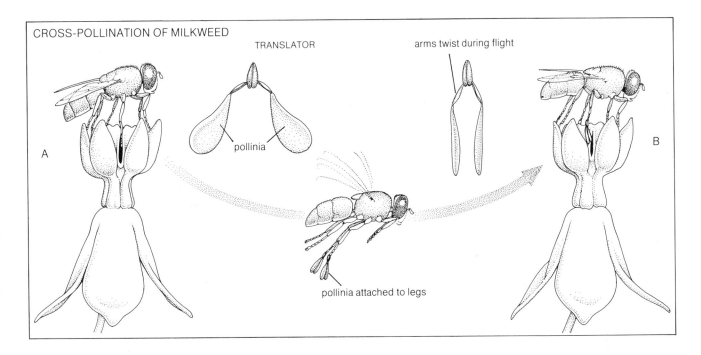

CROSS-POLLINATION OF MILKWEED

TRANSLATOR

arms twist during flight

pollinia

pollinia attached to legs

Above The central column of the milkweed flower (*Asclepias*) is broad and flat-topped. The anthers are fused into a ring with vertical slits between. At the top of each slit there is a clasp, the translator, which is attached to the pollinia. A foraging insect may insert a leg or proboscis into a slit (A) which will become entangled in the clasp. Only a powerful insect species such as a *Heliconius* butterfly (*above left*) will have the strength to pull itself free, dragging with it the pollinia – weaker ones may not escape and will die. Once extracted, the arms of the pollinia automatically twist and the pollen parcels are ready to be deposited on the stigma of another flower (B).

column of the flower which is formed by the fused anthers and the style, and they can only be pulled out by a very special mechanism. In the orchids, the pollinia are kept separate from the receptive stigmatic surface by a sterile projection called a rostellum. Artificial self-pollination works perfectly in practically all orchid species, of which there are about 25,000. Under natural conditions only three per cent of these are self-pollinated.

Separation in time

When the male and female reproductive parts of one flower do not reach maturity at the same time, we can distinguish between protandry, in which the anthers mature and release their pollen before the stigmas are receptive, and protogyny, in which the stigmas become receptive before pollen release. Protandry is found in most composites (Compositae or Asteraceae), and also in many mints (Labiatae or Lamiaceae) and figworts (Scrophulariaceae).

A marvellous example of protandry is provided by the columnar, bumblebee-pollinated, many-flowered inflorescences of foxgloves (*Digitalis purpurea*) and rose-bay willow herb or fireweed (*Epilobium angustifolium*). Both the opening of the flower buds and the production of nectar in the flowers follow a very strict time-pattern, geared to the foraging strategy of the pollinators. The inflorescences of both plants bloom 'from the bottom up' – each flower lasting for several days – which means that the lowest flowers on an inflorescence with many open flowers are female, having gone through the male stage already. These flowers are also the richest in nectar, and foraging bumblebees nearly always visit them first, which means that any pollen deposited on their stigmas is likely to have come from another inflorescence on the same plant or from another plant, but in the usually single-spiked foxglove it will probably have come from another plant. The bumblebees then work their way up the inflorescence, visiting several flowers in succession until the nectar-reward, in terms of calories, is no

As in the case of the rose-bay willow herb, the flowers of the foxglove (*Digitalis purpurea*) are protandrous – the flowers of their single spikes opening from the bottom upwards, being 'male' at first, then 'female'. Bumblebees visit the lower, female flowers first and work their way up to the male flowers before leaving for another spike.

longer commensurate with the energy they have to expend in order to get it. They will then fly off, usually to the lowest flowers of another inflorescence, where the pollen they have picked up from the higher-placed flowers may be deposited on the stigma. The result again is cross-pollination.

Protogyny is very pronounced in a number of families commonly regarded as primitive: water lilies (Nymphaeaceae), the sour-sop family (Annonaceae), the magnolias (Magnoliaceae), arum lilies (Araceae), pipe vines (Aristolochiaceae), and others. Very common plants that demonstrate protogyny beautifully are the plantains (Plantaginaceae). Their many-flowered, spike-like inflorescences, which maintain a position close to the vertical, flower from the bottom up, just as we saw in foxglove and fireweed. Therefore, in an inflorescence with many open flowers it is always the bottom ones that are *male*, while those higher up are *female*. Since plantains depend on wind pollination to a considerable extent, the likelihood of pollen travelling upwards is very low, and the situation therefore is a very good defence against self-pollination.

Unisex flowers

A number of plants have clearly recognizable sex chromosomes, a feature which they share with humans. At first sight one might therefore expect that in general sex in plants is well defined and rigidly controlled by genetics. However, it turns out that there are numerous exceptions. Sex-reversal *is* possible! Sex-expression in plants is therefore fast becoming a focus of scientific attention, and small wonder, for the manner in which it is controlled by the environment has an almost scintillating elegance. Again, flexibility is the key to success. The survival value of having particular sex-distributions in a plant population at a particular time and under particular circumstances (which, of course, may change not only with time but also from one locality to another) is becoming clearer all the time.

Whether rigidly fixed or not, the manner in which in nature the sexual functions are distributed on one plant, or over different plants in a popu-

Inflorescences of the hoary plantain (*Plantago media*) exhibit a condition known as protogyny. When the flowers first open they are in the female stage, later becoming male. In an inflorescence containing flowers in both stages, the upper ones will be in the female stage, making it unlikely that they will receive wind-borne pollen from the male flowers below.

44

The showy lamb's-tail catkins of hazel (*Corylus avellana*) are a familiar sight in spring – these bear the male flowers and each may shed up to four million pollen grains as they bob in the wind. One might be forgiven for not noticing the separate scaly female flowers from which protrude numerous crimson-red stigmas (*top left of photograph*).

lation, varies widely. When staminate and carpellate ('male' and 'female') flowers are formed on the same plant the condition is monoecious (the word refers to 'one household', from the Greek word *oikos*). When they are on different plants of the same species, that species is said to be dioecious. Probably the best-known dioecious family in northern temperate regions is that of the willows and poplars (Salicaceae). Good examples of monoecious plants are birches (*Betula*), alders (*Alnus*), hazel (*Corylus*), cattails (*Typha*), most sedges (*Carex*), and many Arum lilies (Araceae). In general, monoecism reduces the chance of inbreeding, and not surprisingly some species in this category combine both monoecism and dichogamy, which further reduces that chance.

Apart from enhancing outbreeding, monoecism offers another advantage. Once the pollen-dispersing function and the pollen-capturing function of a flower are separated, further developments may lead to a degree of specialization and perfection of these functions, which might be hard to achieve in a bisexual or hermaphroditic one. For example, the male inflorescences of certain oak species, although they hang down, appear above the bulk of the canopy, and are long as well as flexuous. These features certainly optimize the dispersal of pollen. The female flowers, in contrast, do not have too exposed a position, but are firmly attached to a branch, so that the fruits which develop later will not fall off the mother plant prematurely. Each pistil has a wide-spreading but delicate stigma, which makes it a marvellous pollen receptor.

Recently (1981), scientists have looked at the number of male and female flowers on plants of three monoecious species in different localities in Utah, some on very dry mid-slope sites and some on sites with more soil moisture such as bottomland and stream-sides. It turned out that the proportion of male flowers is higher on dry sites, while on sites with more moisture, female flowers are over-represented. For a number of dioecious plant species in Utah, the situation is very similar in that dry sites are characterized by a predominance of male plants. Offhand, one could think of two possible explanations for this. Perhaps the male and female plants, when young, are uniformly distributed in the landscape and occur in roughly equal numbers, but the females are 'weeded out' in the less favourable sites. It is perhaps more likely, however, that all the plants have the potential of being either male or female, and that the site, with its particular physical circumstances, determines to which sex they will belong.

Many investigators have found that environmental stress tends to incline the sex-ratio towards maleness. One of the longest-known and most spectacular cases may well be that of jack-in-the-pulpit (*Arisaema triphyllum*) and some of its close relatives, such as *Arisaema dracontium*. In the first year of its existence, a jack-in-the-pulpit plant is usually male, in the second (when it is larger and well established) it is female. By growing a plant under dry, poor conditions, one can *make* it male; by giving it a very rich environment, with frequent applications of manure, potassium fertilizer, and so on, one can promote femaleness. Robbing a plant of most of its reserve material by cutting off parts of the corm will also lead to maleness. As a strategy for survival of the species, the sex-change of jack-in-the-pulpit makes excellent sense. In a poor environment, the production of seeds and fruits by a plant might well prove too much of a burden, and the plant might perish. On the

45

other hand, if it 'decides' to be male, it can make pollen, which in terms of energy expenditure is relatively cheap, and can still make a contribution to the continuation of its species.

Self-incompatibility

In the USA, an unintentional large-scale demonstration of the bad effects of self-pollination took place in the early years of the present century when large orchards of a single variety of apples, pears, sweet cherries and plums were started. Young orchard trees are usually obtained by vegetative propagation, and in many cases all the trees in one orchard therefore formed a clone of individuals that were genetically identical. When the trees reached maturity several years after planting, very few, if any, fruits were produced. Each tree did produce both pollen and ovules, but the sperm cells in the pollen tubes failed to reach the egg cells in the ovules because the tubes could not grow properly through the stigmas and styles of the pistils.

The trouble lay with the plants, and was *genetic*. Having been obtained vegetatively from the same stock, the commercial orchard trees of one variety are like identical twins; they all have exactly the same genetic constitution, and it has been demonstrated repeatedly that there is an inhibitory effect of a pollen-receiving (or 'host') pistil on pollen-grains (or tubes) from genetically identical flowers. The negative effect can make itself felt in different places. In some plant species it is the germination of the pollen grains that is blocked; or, if they germinate at all, the emerging pollen tubes are not allowed to enter the tissues of the stigma. In other species the pollen tubes, having travelled through the stigma, are prevented from entering the styles. In still other plants, downward growth of the tubes towards the ovules is slowed down to a snail's pace, so that the flower's life is over before the tubes can arrive at the ovules. Even if they get that far, an insurmountable 'road block' may prevent them from entering the ovules even at that late stage.

The multi-storey flower

The final method of promoting outbreeding that we shall consider is called heterostyly – the development, in one plant species, of two or even three types of bisexual flowers, in which anthers and stigmas are found at different levels.

Charles de l'Ecluse (or Clusius), the French refugee who in the sixteenth century started the Hortus Botanicus in Leiden, was probably the first person to notice that the flowers of certain primrose species (*Primula*) come in two types, and that these are borne on different plants. One type, called the long-styled or 'pin' form in England, has long styles and short stamens, while the other, the short-styled or 'thrum' form, has short styles and long stamens. The level of the stigma in pin flowers corresponds with that of the anthers in thrum flowers, and vice versa. Nowadays, plant species that display this condition are called distylous. So far, they have been found in only 24 families of plants, and in most of these in only a few genera. For unknown reasons, the coffee family (Rubiaceae) has more distylous species than all other families combined.

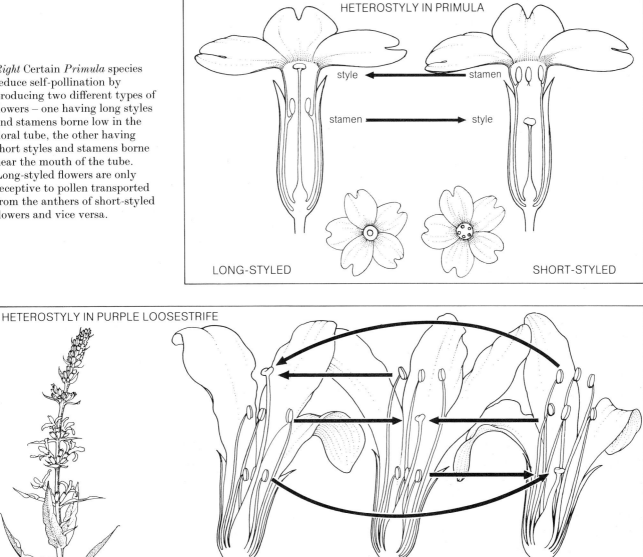

HETEROSTYLY IN PRIMULA

style ← stamen

stamen → style

LONG-STYLED

SHORT-STYLED

HETEROSTYLY IN PURPLE LOOSESTRIFE

flower spike

three different levels of anthers and stigmas

Right Certain *Primula* species reduce self-pollination by producing two different types of flowers – one having long styles and stamens borne low in the floral tube, the other having short styles and stamens borne near the mouth of the tube. Long-styled flowers are only receptive to pollen transported from the anthers of short-styled flowers and vice versa.

Purple loosestrife (*Lythrum salicaria*) possesses twelve stamens, organized in two groups of six, which come in three sizes – long, medium or short. Each flower contains stamens of two lengths only. The single pistil can also be long, medium or short and is always of a different length to the stamens. Each plant possesses flowers with only one of the three possible combinations of stamen and style length. Cross-pollination is enhanced because a flower is unlikely to receive pollen from another flower of the same type.

In 1877, Charles Darwin published a wonderful book on these plants under the title *The Different Forms of Flowers on Plants of the same Species*. The different positions of stamens and pistil in the two flower-forms of the common primrose, he argued, lead to pollen deposition on different parts of the bodies of pollinating insects. The parts of a pollinator carrying pin pollen would of course have a very good chance of coming into contact with a thrum stigma, and vice versa. Thus outbreeding would be promoted. Darwin also found that 'legitimate' pollination – of a pin flower by thrum pollen, or vice versa – results in a much higher degree of fertility than 'illegitimate' pollination of a stigma by pollen of a flower of the same type. Clearly, the 'mechanical' distyly system operates in combination with a chemical system which also 'discourages' self-pollination in a population. In a way, distyly is a disadvantage, the reason being that if the number of pin and thrum flowers is about the same, a given plant can 'interact' with only one-half of the plant individuals in a population, in contrast to what we see in most other plant species. It is, therefore, not easy to explain how the distylous condition was brought about in the process of evolution! It remains a matter of much debate.

CHAPTER FOUR

Adaptation and co-evolution

It is too bad that often there is such a wide gap between common parlance and scientific terminology. Many biology students, for instance, are initially confused by the word 'adaptation'. When we say that a certain person adapts beautifully to adverse circumstances, we are obviously referring to a short-range, dynamic process – something that happens in the life-time of the person involved. In a scientific context 'adaptation' is most commonly used with a strong *static* connotation. We notice, for instance, that dolphins and whales function exceedingly well in the oceans, where they spend their whole life and produce their young. Their organs are 'just right' for the job, which means that there is a beautiful 'fit' between the organism and its environment. But nobody would ever dream of suggesting that this was brought about by the individual's conscious effort. Yet we call them 'adapted'. The dynamism comes in when we consider how the situation developed – probably through an extremely slow, gradual change from generation to generation; essentially a combination of genetic processes and natural selection.

In pollination biology we frequently consider pairs of organisms – one plant, the other animal – that are adapted *to each other*. We notice, for instance, that the flowers of certain evening primroses have such a long, narrow corolla tube that only hawkmoths – animals with an extremely slender, long and flexible proboscis – can get at the nectar hidden in that tube, pollinating the plant in the process. Obviously, both partners benefit: the plant can now produce offspring, and the hawkmoth receives food to fulfil its energy requirements. Among biologists, there is now a reasonable consensus that situations of this type were brought about in a slow process of 'co-evolution'. Where a layperson, looking at the here-and-now, might come up with the question, 'Which of the two came first, the long corolla tube or the long proboscis?', the biologist realizes that the answer to a query of this sort is usually that neither came first: the features have evolved side by side.

Extreme cases of co-adaptation between flowers and their pollinators are impressive, but it is as well to bear in mind that few flowers are so specialized that they can be pollinated by only one pollen-carrier. One has only to look at the flat, open inflorescences of daisies, or at those of wild carrot, to realize that many plants follow an 'open house' policy and roll out the welcome mat for all comers. It is true that the general evolutionary tendency in pollination has been one leading towards greater refinement and specialization, creating a very strong interdependence between flower and pollinator. However, there is a built-in danger here. A 'specialized' system may continue to work supremely well over long periods, sometimes over millions of years.

A number of adaptive features are evident in the relationship between hawkmoths (here the elephant hawkmoth) and many honeysuckles. Nectar is available deep in the flower's narrow tube, reachable only by a very long proboscis; also, as most hawkmoths are nocturnal, the flowers are fragrant at night and are pale in colour.

However, if for some reason one of the partners were to be eliminated, which might be the result of something as simple as a change in climate, the other would automatically be doomed unless it could quickly develop some compensatory mechanism such as self-pollination. Extreme pollinator-specificity can also be a barrier to the successful colonization, by a plant, of new areas. Any such move would, of necessity, require the simultaneous introduction to those areas of the pollinator. It is no coincidence that the most successful plant colonizers of all – those we call weeds – are, in general, characterized by very unspecialized breeding systems.

A fact frequently overlooked is that adaptation displays both a positive face *and* a negative face. On the one hand, a flower will attract or encourage certain pollinators; on the other, it will discriminate against others, sometimes with complete success. For instance, the pure scarlet colour of a certain flower may act as a strong attractant for hummingbirds, while at the same time discouraging most insects which, lacking red vision, simply see it as black. An evening primrose or honeysuckle flower, open and fragrant only at night, will automatically favour nocturnal hawkmoths while discriminating against sun-loving creatures such as butterflies and honey-bees.

It is interesting to notice that discriminatory measures in flowers can sometimes be too strong, the result being that they will boomerang. For instance, the sturdy flowers of *Thunbergia grandiflora*, with their very narrow flower entrances, are so hard to get into that they almost invite burglary – even by the strong carpenter bees that are their legitimate pollinators. These matters will be discussed in more detail in Chapter Five. In this chapter we will examine the co-evolution of certain flowers and their specific pollinators, which has led to the creation of special 'flower classes'.

The open-house principle and the role of flies and beetles

In spite of what we have just said about co-evolution and the resulting close fit between flowers and their pollinators, it is a fact that many plants are generalists offering pollinating animals an 'open house' invitation in the form of easily accessible flowers with freely exposed nectar and/or pollen. Often, these flowers are grouped together in large, showy, flat or gently rounded inflorescences. Good examples are found in the parsley family (the Umbelliferae) with such plants as cow parsley (*Angelica*) and Queen Anne's lace, or wild carrot (*Daucus carota*); the daisy family (Asteraceae or Compositae); the ivy family (Araliaceae); and the honeysuckle family (Caprifoliaceae) with such plants as elderberry (*Sambucus*) and *Viburnum*. On their inflorescences one can usually find a motley crowd of insects – bees, wasps, flies and beetles of many kinds, and even some butterflies, although these are generally considered specialists.

Flies and beetles are very ancient groups, containing an almost incredible number of species: their diversification is equally impressive. In the realm of floral biology, therefore, it does not make sense to lump all the flies together in one category and all the beetles in another. There is, after all, little in common between the dashing long-tongued bee-fly (*Bombylius*), which in its life-style reminds one of a hummingbird, and a lowly carrion- or dung-fly. One generalization that can safely be made is that there are a number of flies and beetles that are not really adapted to flowers at all, but which are

Hoverflies belonging to the family Syrphidae, such as this *Syrphus balteatus*, have a relatively short proboscis and visit flowers with easily accessible nectar such as are common in the carrot family. Their markings mimic those of bees and wasps.

nevertheless essential to the reproductive processes of certain plants. These are the carrion, dung and mushroom flies or beetles that are trapped by various flowers or inflorescences such as those of Dutchman's pipe (*Aristolochia*), lords and ladies (*Arum maculatum*) and voodoo lilies (*Sauromatum guttatum*).

Syrphid fly pollinators

Of the flies that are more than just hapless victims of trap-flowers, the syrphids deserve special mention. Many of them are colourful creatures, mimicking wasps in their appearance yet perfectly harmless. Their role has definitely been underestimated in the past, and some investigators have even considered the smaller syrphids in the same light as ants; they were thought of as pollen-thieves, which, because of their small size, would fail to bring about pollination. However, it has recently been shown that in certain areas, for instance on the subalpine meadows of Mount Rainier in the American Pacific Northwest, they are extremely effective pollinators which develop flower constancy. In the Netherlands and some other parts of Europe, syrphid flies in the *Melanostoma-Platycheirus* group have been shown to be involved in a specific relationship with certain plantains (*Plantago*), sedges (Cyperaceae) and cattails (*Typha*).

Pollination by mushroom-gnats

In the late years of the nineteenth century, the highly esteemed Italian botanist Arcangeli described a pollination case so 'unbelievable' that even his friends began to fear that he had fallen prey to early senility. It concerned the curious little arum lily *Arisarum proboscideum*, known as the mousetail plant. Its inflorescence has a cylindrical, vertical, but slightly bent-over floral chamber that is completely closed except for an elliptical window that looks slightly earthwards. On top of the floral chamber, and forming an extension of it, there is a dark-coloured, extremely slender, drawn-out and curved tip, the 'mousetail'. (In spring, the plants bear inflorescences and leaves at the same time, and when one has a whole dense bed of them in the garden, they do indeed give the impression that a small army of mice – all sticking up their tails at the same time – has found refuge among the foliage.)

A small flying insect, coming up from the forest floor and entering the floral chamber through the window, is immediately confronted by the appendix of the inflorescence, which in this case is not hard and smooth as it is in many other arum lilies but is spongy and full of little depressions. The organ is also off-white in colour so that the overall visual impression it gives is deceptively like that of the underside of the cap of a *Boletus* mushroom. Arcangeli claimed that the plants' pollinators were female fungus-gnats – animals that normally breed in decaying mushrooms! The mousetail plant fools them so successfully that the females deposit their eggs in, or on, the appendix. Getting down more deeply into the floral chamber they will also pollinate the plant. It is gratifying that further research in the 1960s and 1970s exonerated Arcangeli completely. The 'unbelievable' story was true.

Fungus mimicry is a fairly widespread pollination strategy and in most cases the pollinating gnats lay eggs that are bound to perish. Most of these fungus mimics are forest-dwellers: the plants remain close to the ground and produce flowers that are dark purple or brown in colour, with pale or translucent patterns. To the human nose at least they are either scentless

The inflorescence of the Mediterranean mousetail plant (*Arisarum proboscideum*) is known to attract female mushroom-gnats, which are deceived into 'thinking' that the mushroom-mimicking spadix concealed within its chamber is a suitable egg-laying site. The function of the long, narrow 'tail' is not fully understood, but such appendages do seem to be common to flowers pollinated by small flies.

Above Guided by linear markings on the 'tails', a fungus-gnat has been lured to the fleshy petal of this wild ginger (*Asarum caudatum*). She will lay her eggs in the throat of the flower.

Right Native to Colombia in tropical America, the orchid *Masdevallia bella* has a gill-like blade and a fleshy claw on its lip petal which form an almost exact replica of the underside of a mushroom.

53

or musky in odour. Usually the flowers are simple urn- or kettle-shaped traps containing structures that closely resemble the lamellae or pores of mushrooms and toadstools. Another element in their fungus mimicry is the intense local transpiration displayed by these structures during the period when the flower is active. Fungus gnats of both sexes are involved in the pollination and are misled by a combination of fungus-like features – odour, colour, shape, texture and humidity.

In the American wild ginger (*Asarum caudatum*) and other *Asarum* species, the female fungus-gnat deposits her eggs in the throat of the flower. The larvae that hatch from them start eating almost immediately, but the tissues of the flower are so poisonous that the young animals die very quickly. It has been found that in some localities 35 per cent of the flowers of the American wild ginger contain eggs and larvae. This, then, is a case of 'ruthless' exploitation of a pollinator by a plant, comparable to the cases of *Stapelia* and certain water lilies which we will describe in a later chapter.

Probably the most spectacular cases of mushroom-mimicry are found in some *Masdevallia* orchids. Here, the central part of the flower displays a number of fleshy 'gills' radiating out from one point and looking deceptively like those found on the underside of fairy-ring mushroom caps.

Beetle-traps

Although it might be quite logical to observe that 'a trap is a trap is a trap', it is a fact that some trap-flowers (and inflorescences) are designed to serve beetles only. Their pollination stories are amazingly similar, even though they may belong to different families and so exhibit differences in the construction of their trap. It may be significant that the families with flowers or inflorescences that trap beetles (possibly the first pollinators in history) are themselves 'primitive', such as the water lilies and the arum lilies. Their flowers attract the beetles by producing strong odours, sometimes fragrant but more often carrion- or dung-like. Often, they generate heat to evaporate the odoriferous compounds, so making them more effective. They often provide the beetles with special food-bodies or nutritious tissues. And finally, since the visiting beetles, carrying pollen, must first be imprisoned to 'serve' the receptive pistils, and then be released later to export pollen from the prison to other receptive flowers, the timing mechanisms of beetle-traps must work with clock-like precision. Here, we will look at just a few representative cases.

The flower buds of the famous *Victoria amazonica* from Brazil, with its huge round floating leaves with upturned rims, open shortly after dark; the lovely creamy white flowers are warm at that time, and emit the sweet fragrance of tropical fruit. The next morning, the flowers are closed again. In late afternoon they open once more, but the colour now is pinkish or purplish while the odour is almost gone. In *Victoria*'s native environment, the fragrant fresh flowers (in which the pistils are receptive) will attract large numbers of good-sized *Cyclocephala* beetles, related to our June bugs and Japanese beetles. At least some of them may carry *Victoria* pollen, picked up in flowers visited earlier. While locked up during the night, the voracious beetles may do quite a bit of damage to the interior of their prison. However, they also pollinate, and enough ovules survive their onslaught to

The floral spadix of this arum lily, *Xanthosoma*, attracts *Cyclocephala* beetles. A plant sometimes cultivated for its edible tubers, it is here photographed in the cloud forests of Costa Rica.

produce viable seeds. Pollen will not be shed until the next afternoon or evening, when the flowers open again. The released prisoners, carrying the pollen, will be trapped again by receptive *Victoria* flowers, and so the story is repeated.

In the East Indies, the common arum lily *Amorphophallus variabilis* is known as kembang bangke or carrion-flower. Its powerful odour has over-tones of a sickening sweetness, and it is therefore not too surprising that it is pollinated by nitidulid flower beetles. The inflorescences always begin to smell at 4.30 in the afternoon. Its huge relative *Amorphophallus titanum* from the jungles of Sumatra has inflorescences that may reach a height of 2.5 to 3 metres (8 to 10 ft). These produce considerable heat, and their carrion smell is quite overpowering. The flowers are pollinated by large *Diamesus* beetles that are held prisoner in the floral chamber for several days but are well fed there. Not quite as large, but still impressive, is the American *Xanthosoma*. In Costa Rica its pollinators were again found to be *Cyclocephala* beetles. The animals use the inflorescence as a mating arena, possibly activated by the heat that is produced there. The Mediterranean species *Dracunculus vulgaris* (often grown in American and western Euro-pean gardens) is known as black lily of the Nile because the inside of the spathe of the inflorescence has a beautiful purplish-black colour. It possesses the largest flowers, or rather inflorescences, of the European flora; 30 to 60 centimetres (1 to 2 ft) is average, but some may be as tall as 1 metre (3 ft). Two *Dracunculus* plants in Seattle yielded a total of 298 beetles belonging to fifteen different species. All of these were of the type usually found on excre-ment and carrion, with the exception of two that were compost beetles.

As to the beetles found on more modern, open flowers, it is fair to claim that some of them have become highly adapted pollinators. Members of the cockchafer family, such as June beetles, often found on wild roses, spiraeas, etc., are pollen-eaters, while some longhorn beetles (Cerambycidae) have become nectar-feeders with a slender head sticking forward in the same direction as the body, and mouthparts that are somewhat brush-like and well suited to lapping up fluids. One of the most extreme cases of specializ-ation is found in the beetle genus *Nemognatha* (literally: thread-jaw), in the family of the blister-beetles. Here, the maxillae, or lower jaws, are developed to such an extent that in many species they are longer than the body, or rather, each maxilla bears a long, almost straight, sabre-like appendage. The inner hair-covered margins of the sabres touch each other, so that a sort of gutter is formed through which nectar can move to the beetle's mouth. It can legitimately be compared with the proboscis of a butterfly since the mouthparts that form the structure are the same in both cases. However, the extended instrument of *Nemognatha* cannot be coiled and uncoiled like that of the butterfly.

In southeastern Europe, true chestnut (*Castanea*) shows considerable adaptation to beetle pollination. The male flowers have a strong, somewhat fishy odour, and although they attract all sorts of insects, there are places (Carinthia, for instance) where more than half the visiting insect species (53 out of a total of 103) are beetles. As long as it is not too old, the pollen is sticky, which favours transport by insects. Towards the end of the flowering season, however, the pollen gradually loses its stickiness and the wind becomes the pollinating agent.

55

Pollination by bees

There can be no doubt that on a global scale, bees (and especially honey-bees and bumblebees) are the most important pollinators. It has been calculated that in pre-war Germany alone, honey-bees pollinated about ten trillion flowers in the course of a single summer's day. To make 0.5 kilogrammes (1 lb) of honey, bees have to visit about ten million clover blossoms. This leads to the production of about 13.5 kilogrammes (30 lb) of clover seed – enough to seed 1.2 hectares (3 acres) of land! The pollen collected by a single honey-bee colony may be more than 29 kilogrammes (64 lb) in a year.

In North America there are about 5,000 species of bee, 100 different species having been counted on the single crop alfalfa, a plant which – with its high protein content – has become one of the most important American crop-plants. In Eurasia, honey-bees have been associated with man from time immemorial. They probably originated in South-East Asia, but European travellers carried them to the Americas, South Africa and Australia with the result that there also they have begun to dominate the pollination scene at the expense, very often, of the original pollinators.

Earlier in this book we have considered the honey-bee's colour vision, sense of smell, flower constancy and time sense. It now remains for us to deal with their superb communication system, and to compare their social organization and foraging strategies with those of the more individualistic bumblebees (*see* p 62).

The lowest sepal of the flowers of the Himalayan balsam or policeman's helmet (*Impatiens glandulifera*) is petaloid and helmet-shaped, providing the rather heavy bumblebee with a platform on which to stand while probing into the recurved tail-like spur in search of nectar.

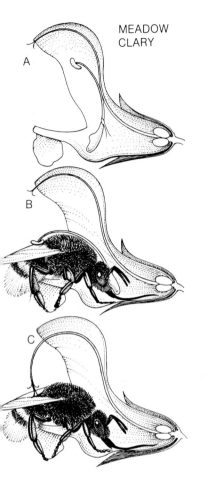

MEADOW
CLARY

A

B

C

When a flower of the meadow clary (*Salvia pratensis*) first opens it is for all practical purposes male (A) – the pollen is ripe, but the stigma lobes are unreceptive. A bumblebee climbing into the flower in search of nectar activates a pivot mechanism (B), bringing a pollen-laden anther down on to his back. When she leaves, the anther springs back. During the next few days the flower becomes female – the pistil stretching and curving downwards, and the stigma lobes spreading out and becoming receptive. A visiting bee on this occasion must touch the stigma (C), dusting it with pollen if she has just visited a male flower.

Typical bee-flowers are open in the daytime, have a minty fragrance, and offer their visitors nectar, pollen or both. They are brightly coloured (often yellow or blue) in order to attract insects from a distance, while for close-range interaction with their visitors they often possess nectar-guides in the form of detailed colour or smell patterns, which may include special UV reflection patterns. Many species provide the bees with a landing platform, usually in the form of a broad lower lip on which the visitor can alight before pushing its way deeper into the flower. Thus they display bilateral rather than radial symmetry. Efficiency, for plant and animal alike, is served by the fact that the flowers are normally hermaphroditic: during a single visit, a flower both delivers and receives pollen, while the visitors save energy by not having to travel from one flower to another empty-handed. A single pollination visit thus suffices to do the job. In fact, there are many bee-flowers in which the pollination mechanism is such that more than one meaningful visit is not possible. Different bee-flowers deposit the pollen on different parts of the visitors' bodies: sometimes on their back as in true sage and other members of the mint family (nototribic pollination); sometimes on the underside of their body as in vetches (sternotribic pollination). The pollen grains too show their adaptation to insects (bees) in that they are sticky, spiny or highly sculptured, so that they cling easily to the mouth-parts, legs or bodies of the visitors. Since one animal may carry thousands of pollen grains, it is not really surprising that the ovary of the receiving flower contains a great many ovules.

Good examples of bee-flowers are found especially in the mint family (Labiatae) and the legume family (Leguminosae or Fabaceae); we may mention true sages (*Salvia*; except for the scarlet ones that are hummingbird-pollinated), clovers (*Trifolium*) and lupins (*Lupinus*). More specifically adapted to bumblebees are foxglove (*Digitalis*), fireweed (*Epilobium*), monkshood (*Aconitum*), most larkspurs (*Delphinium*), snapdragon (*Antirrhinum*), toadflax or butter-and-eggs (*Linaria*), blueberries and huckleberries (*Vaccinium*), dead nettle (*Lamium album*), lousewort (*Pedicularis*), yellow-rattle (*Rhinanthus*), and many others. Excellent pollen-flowers that are bee-pollinated are European poppies (*Papaver*). The red ones – like the others, probably – reflect ultraviolet and are therefore seen as 'coloured' even by the red-blind bees. They are frequently visited by honey-bees.

There are 400 genera of plants, in 65 families, that have anthers opening through a pore (or pores). In many cases the flowers face downwards. The pollen, which is powdery, has to be shaken out in a very special dramatic process known as 'buzzing', which is practised by bees in several different genera (but not by honey-bees, leaf-cutters and andrenids). Buzz-pollination is characterized by very fast wing-muscle vibrations of up to 300 cycles per second. Splendid examples of flowers built for buzzing are shooting-stars (*Dodecatheon*), European bittersweet (*Solanum dulcamara*) and tomato (*Solanum lycopersicum*).

Pollination by euglossine bees

Several orchid species have evolved flowers that attract male euglossine bees, providing them with oily, fragrant substances, used in the bees' courtship rituals, in return for securing their services as pollinators. The tribe

Euglossini, of the family Apidae, contains five genera and approximately 180 species. All are found in tropical America. The bees themselves are fascinating. Both sexes are equipped with extraordinarily long tongues, which are held along the underside of the body when not in use, and in some species jut out well beyond the end of the abdomen. The females are solitary, and while they collect nectar, oils and pollen from various flowers, it is the males that have exerted the greatest impact on floral evolution. Indeed, among the orchids pollinated by euglossine bees are found the most bizarre and complex flowers in the entire plant kingdom.

The attractants evolved by these orchids are blends of aromatic compounds, each species of orchid liberating a blend that attracts only one, or in some cases, two, species of euglossine bee. The rewards offered to the male bees are waxy or oily substances which the bees subsequently use, after chemical modification, as components of the pheromones used in their courtship rituals. No one has suggested that the bees cannot breed successfully without the orchid substances, for they seem to manufacture their pheromones from substances gathered from a variety of sources – including flowers, fungi and tree-trunks – but the energy they devote to visiting their orchids certainly indicates that this particular source is important to them. Perhaps it gives them an 'edge' on the competition.

The pollination techniques displayed by the different orchid species vary enormously – from simple gullet flowers to flap-trap, hinge and fall-through mechanisms. Perhaps most impressive of all is the pitfall trap of the genus *Coryanthes* – the bucket orchids.

There are about 20 species of *Coryanthes*, all found in tropical Central America. Each one is pollinated by its own species of euglossine bee, as far as current research can tell, and all have the same floral design. *Coryanthes* orchids are epiphytes. They tend to grow in association with ants, which protect the plant and especially the developing buds from attack by herbivores (grasshoppers, caterpillars, etc.) and whose detritus-filled nests provide the *Coryanthes* roots with nutrients. The flower buds develop on long, drooping stems so that the flowers, when they open, hang down below the branch on which the plant is growing. Each stem bears from two to about seven flowers, which open very quickly shortly before, or at, dawn.

The construction of the flower is nothing short of astounding, even to the hardened botanist. The sepals, which bend back like crumpled butterfly wings, reveal a structure that looks like a steep-sided, open-topped bucket, with a mushroom-shaped object growing out of the rim. Over the centre of the bucket hang two rounded knobs, which are in fact glands. These produce a continuous secretion of clear fluid that drips down into the bucket below for the first couple of hours after the flowers open, by which time the flower contains about 6 millimetres ($\frac{1}{4}$ in) of fluid. The glands are then 'turned off'. A short tunnel, rather like the spout on a watering can, opens from the side of the bucket, just above the water line. At the entrance to the tunnel, and partly submerged in the fluid, is a small bump on the wall of the bucket. In the ceiling of the exit end of the funnel are the flower's pollinia and stigma.

The flowers are coloured yellow, or yellow mottled with brown and orange, depending on the species. An observer does not usually have to wait long to find out how this elaborate floral structure works. Within minutes of opening, the flower emits a heavy, soapy and sweet scent. Male euglossine

While collecting an oily secretion from a bucket orchid (*Coryanthes* sp), male euglossine bees often lose their grip and tumble into the 'bucket'. Soaked with fluid the bee can find only one escape route – through a conveniently positioned tunnel in the side of the chamber. However, at the end of the tunnel the flower restrains him, sticks two pollen packages on his back and holds him until the glue has set. Once freed the bee may visit another bucket orchid and the whole process is virtually repeated, but this time a hook at the end of the tunnel neatly removes the pollen packages from his back.

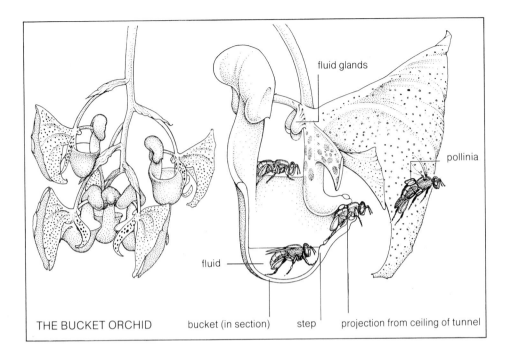

THE BUCKET ORCHID — bucket (in section) — step — projection from ceiling of tunnel — fluid glands — pollinia — fluid

bees, which are usually somewhat rare and secretive creatures, materialize as if by magic, sometimes in their dozens. The bees that visit *Coryanthes* are among the most beautiful of all insects, being varying shades of brilliantly iridescent green, blue and bronze, depending upon the species.

To begin with, the bees do not land on the flowers but simply fly about in great excitement, occasionally ramming one another in mid-air as if competing for ownership of what the orchid has to offer. After a few minutes they start to land on the rim and outer walls of the bucket, and on the mushroom-shaped structure that stands on the rim opposite the two glands. Each bee scrapes frantically at the waxy surface of the flower for a few seconds with its brush-like front feet, then backs off, hovering in mid-air while it transfers the oily substance it has collected to hollow swellings on its hind legs. The bee then lands again, repeating the whole sequence.

With up to a dozen or so bees busily gathering orchid substances from each flower, and indulging in occasional bouts of aerial combat at the same time, it is not surprising that sooner or later a bee will lose its footing and tumble into the centre of the flower. It immediately sinks into the clear fluid (experiments have shown that the fluid contains a wetting agent, and consequently has a very weak surface tension). The bee struggles to escape, but the walls are waxy and smooth and it can make no progress. It would quickly drown in the orchid's fluid but for an escape route built into this floral trap – the tunnel. As the bee thrashes about in desperation, its front feet make contact with the bump on the side of the bucket. The bump is a sort of step-ladder, leading the waterlogged bee into the entrance of the tunnel. The tunnel is an extremely tight fit and the bee literally forces his way towards the exit. At the very moment that he seems to be gaining his freedom, a projection from the ceiling of the tunnel locks down into the gap between his thorax and abdomen. Despite his energetic struggling he is held fast, while the flower securely glues its pollinia on to the bee's back! What determines the length of time the bee is held captive, how the mechanism is

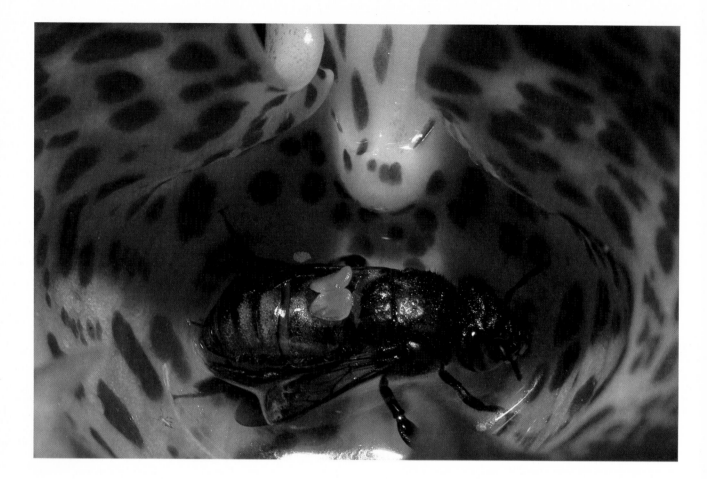

activated, and how, when it is terminated, the bee is finally released, is not known. After about ten minutes the pollinia-laden bee finally emerges, and flies off to dry.

Meanwhile, what is happening to the other bees, busily gathering their orchid substance? Many of them suffer the same fate as the first bee, only now that the pollinia have been removed the bees simply pass through the escape tunnel unhindered, like spectators through a turnstile. Bees continue to be attracted until the flower is pollinated, or until the flower wilts at the end of its second day, whichever is the sooner. Pollination occurs when, and if, a bee bearing pollinia from another *Coryanthes* plant tumbles into the flower. When this bee makes his escape, a pick-up hook on the ceiling of the tunnel plucks the pollinia from his back as he forces his way past. The chances of this ever happening, though, must be slim. *Coryanthes* is not a common orchid, and it flowers irregularly. The bees are not exactly common either, and although a *Coryanthes* in bloom is said to attract male bees from up to five miles away, few if any of them will be bearing pollinia from another plant. What is more, by no means every visitor falls into the bucket. The dice seem to be loaded against the orchid. Observation of *Coryanthes* in the wild rarely yielded plants with more than one pod per flower stalk, and in most cases none at all. However, in the lucky event of pollination occurring, the resultant seed pod contains hundreds of thousands of microscopic seeds, which drift off on the wind when the mature pod eventually splits open and frees them.

Above A pair of pollinia are securely cemented to the abdomen of this euglossine bee – a payload acquired during a previous encounter with a bucket orchid (*Coryanthes* sp). During its escape from this second flower's fluid-filled trap, the bee will have the pollen parcels plucked from its back, thus successfully cross-pollinating the orchid.

Euglossine bees and the *Catasetum* sharpshooter

The sharpshooter principle of pollination is employed by the many South American orchid species that make up the genus *Catasetum*. But the story is further complicated in that only the male flowers employ the technique: *Catasetum* is one of the few orchids in which one finds separate male and female flowers – usually on different plants. They are so different in structure that it is hard to believe that they belong to one and the same species. Rather drab in their appearance, both the male and female flowers produce a strong fragrance which, in the case of the male, puts in its appearance two to three days after the flowers open. In the female, the fragrance comes several days later.

The pollinators are male euglossine bees, in this case *Euglossa* species, keen on collecting the perfume that is produced by the glandular tissue in the flowers' lip cavity. When the animals penetrate into the odour-chamber of the male flowers, they can hardly fail to touch one, or both, of the sharp tips of a most peculiar pair of appendages, known as the 'horns' or 'antennae', which arise from the part of the flower where the pollinia lie hidden. A thin elastic stem connects them with a so-called viscid disc, which is positioned in a little pouch in such a way that the stem is kept in a bent position, under tension. Even a slight bending of the horns, resulting from the gentle touch of a bee, releases the viscid disc from its pouch with formidable force, the reason being that the bent stem of the pollinia snaps up like a metal spring. The whole pollinarium (the combination of the two pollinia with their stem and viscid disc) is catapulted out, goes through a partial backward somersault, and hits the bee, with the viscid disc first. The impact is so strong that

There are many species of vividly coloured euglossine bees, but the males of only one species will find the highly selective blend of perfume emitted by this particular *Gongora* orchid attractive.

the insect is sometimes knocked out of the flower! Experiments have shown that if the bee did not intercept the missile with its body, the pollinarium would cover a distance of 80 centimetres (almost a yard). Small wonder, then, that the viscid disc adheres to the insect immediately and firmly! The marksmanship of these flowers is almost uncanny, but the particular part of the bee's body that will be hit differs from one *Catasetum* species to another.

As soon as the pollinarium has been fired, the odour of the flower begins to change, and after a short time it is gone completely. The job of the male flower is finished. The female flowers do not start producing their scent until several days later, and the scent is not released continuously, but only for a few hours (usually one to two morning hours) on successive days – at least for as long as the flower remains unpollinated, and that may take anything up to a month.

When a male euglossine bee carrying a pollinarium moves into the odour-cavity of a female flower the pollinarium dangles down vertically. In leaving the flower with a load of perfume, the insect moves backwards, and if the *Catasetum* is of the same species that furnished the pollinarium – but *only* in that case – the pollinia will be in exactly the right position to get caught in the groove that represents the receptive tissue of the stigma. Here they will stay, while the stem and the viscid disc (which break off) will be carried off by the bee. The flower has been pollinated. Within a half-hour, it will lose its odour, and from now on not a single bee will pay attention to it.

Each *Catasetum* species has its own specific odour, and will therefore attract its own special *Euglossa* bees. Furthermore, the catapult mechanism works with such extreme precision that the pollinia are always attached to a particular part of the pollinator's body: the 'correct' part for purposes of pollination. The result is that in nature hybridization between *Catasetum* species is exceedingly rare.

The indefatigable honey-bee

Food-gathering by honey-bees – in stark contrast to what we see in bumble-bees, who really are rugged individualists – is a community enterprise, depending on a superb system of communication which Karl von Frisch (who discovered it) has called the 'dance language' of the bees. The animals convey what they have to say by means of movements and scents, and can indicate both the distance and the direction of a rich food-source with great accuracy. It is fair to say that in the animal world this is a unique achievement. Here is how the system works: When a scout bee discovers a rich source of nectar and/or pollen, let us say a group of plants of one species that has just begun to bloom, she will return to the hive, where she performs a so-called waggle-dance, following a path that is shaped somewhat like a figure eight and wagging her tail energetically as she goes through the straight part of the figure. This dance is really an invitation to her hive-mates to fly out and start searching for the food-source. Her excitement is infectious: the hive-mates will begin to follow her dance-movements closely, and will thus first of all notice the scent of the food-source which still clings to her body. So, the animals now know what to look for (or, rather, smell for) once they are close to the source. But of course they have to get there first,

Given precise directions by a scout bee at the hive, a worker honey-bee (*Apis mellifera*) can home in very efficiently on food-sources such as this *Aubrieta*. Honey-bees collect nectar and pollen to furnish the nutritional needs of their colony, which may contain more than 50,000 larvae and adults.

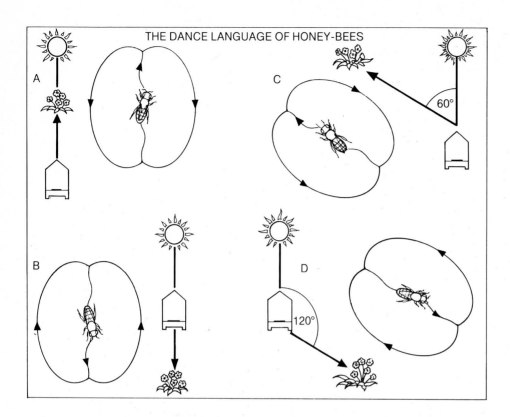

THE DANCE LANGUAGE OF HONEY-BEES

The sophisticated 'waggle dance' performed by scout bees informs fellow workers of the precise whereabouts of a food source. Assuming the top of the page represents the top of a hive, these diagrams show but a few examples of the dance. If the food source is found by flying towards the sun from the hive (A), then the 'waggle' part of the figure-of-eight routine is danced in an upward direction. Diagrams B, C and D show respectively the orientation of the dance if the food is directly away from the sun, at 60° or at 120° from it.

and the dancing bee must also tell them how to do that. She indicates the distance by the tempo of her dance, that is by the number of tail-wags or turns per minute. A high number indicates that the food-source is close by, a low number that it is farther away.

For indicating the direction of the food-source relative to the hive, the dancer uses the sun as a compass. If she dances in the open air on a horizontal plane (admittedly a very exceptional situation), and she moves directly towards the sun as she goes through the straight portion of her dance-path, she is telling her hive-mates that they must fly straight towards the sun after leaving the hive. If the scout bee moves away from the sun during the straight part of her dance, her colleagues must also fly away from the sun to locate their food-source. In most cases, of course, the direction of the food-source forms an angle with the direction of the sun, and this the scout bee indicates by altering the direction of the straight part of her dance by the same angle.

In reality, the bee-dances are usually carried out inside the dark hive, on the vertical surface of a comb, and in this case gravity is used as a reference point. That is, if the straight part of the scout's dance is vertical and up, the message is: fly straight towards the sun. Vertical and down means: move in the opposite direction. Upward and 60 degrees to the left of vertical means that the foraging bees should fly 60 degrees to the left of a line that would connect them and the sun, and so forth. In practice, the system is even better than we have described it here, because the dancer applies subtle corrections if necessary. For instance, when the food-source is very far away, so that considerable time elapses before the returning scout bee can deliver her message to the hive, the sun has changed position by the time the recruits are leaving to find the food-source. The dancer, however, indicates the correct (new) position. She can also make corrections for (say) a head-wind, which makes flying more difficult and thus in effect increases the distance that has to be covered by the bees.

The upshot of all this activity is that a food-source, once found by a scout bee and then advertised by her in the hive, will soon be exploited by a multitude of followers. Of tremendous importance in this is the phenomenon of floral constancy which we have already described briefly. Recall that it takes only a few seconds for a novice bee to form a mental tie between a flower and its particular colour and odour – a tie that becomes practically unbreakable after only one or two flights. What the bees actually learn is the complete 'package' of scent, colour and pattern cues presented by a particular kind of flower, and this in turn is associated with the time of day at which the reward (nectar or pollen) is available. In this manner bees are able to establish a foraging schedule – involving various different food plants – which they will stick to for several days in a row.

In recent years, a great deal of attention has been given to the foraging strategies of bumblebees. It stands to reason that they should not spend more energy on getting food than they get out of it, but they learn fast. It has been found that in freshly emerged workers of *Bombus vagans*, flower constancy develops during the first one or two flights. About half a dozen different kinds of flowers may be probed during the first 50 visits, but the animals soon learn to restrict themselves so that just one or two species serve as hosts. Still, bumblebees maintain a certain amount of interest in a number of different species; just like students, they have 'majors' and 'minors', and when a major flower species peters out as a food-source, they switch to another. Thus, there is a beautiful flexibility in bumblebee behaviour.

Pollination by ants

Until recently, no beneficial role in pollination was assigned to ants; they were considered to be just thieves of nectar and pollen. At best, certain investigators were willing to concede that an 'ant-guard', provided with food by the plant in the form of extra-floral nectar, might keep potential flower-burglars such as carpenter-bees at bay. The discovery, in 1963, that ants are responsible for the pollination of the 'ant-plant', *Orthocarpus pusillus*, was truly revolutionary. Nowadays, it is recognized that there is a genuine syndrome of ant-pollination. It was carefully described in 1974, on the basis of observations on *Polygonum cascadense*, as follows. Since worker ants are non-flying animals that do not expend many calories on travelling, the whole system is 'low-energy'. The nectaries are small and produce a quantity of nectar so modest that larger insects are not interested. The flowers, likewise, are small, sessile, and close to the ground, with minimal visual attractions. They produce only small quantities of sticky pollen grains, so the ants are not forced into intensive self-cleaning activities that would remove the pollen from their bodies. Outbreeding is promoted because on each plant only a few flowers are open at the same time, and also because these lowly plants occur in groups, with their twigs and foliage closely intertwining; the latter feature strongly facilitates movement of the ants from one plant to another.

In all probability, ant-pollination is much more common than previously believed. On Mount Rainier it was observed on *Polygonum newberryi*; it has been found in *Epipactis* orchids (side by side with pollination by other animals); and in Seattle, it probably occurs in low-growing buttercups (*Ranunculus repens*). It is believed that the phenomenon is most often found in hot, dry habitats, which are certainly among those favoured by ants.

Pollination by wasps

The wasps form a large and highly diversified group. The true wasps, Vespidae, are social and nurse their brood. In contrast to honey-bees their larvae are carnivorous; the adults, of necessity, are predators or feed on carrion, and for that reason nectar is important to them only as a source of carbohydrate for their own energy needs. As pollinators, they don't begin to compare with honey-bees. In fact, they may exert a strong negative influence in the area of pollination by victimizing honey-bees, other wild bees and butterflies. The so-called beewolf (*Philanthus triangulum*), a predatory solitary wasp, specializes on honey-bees and since, in some years, it may occur in large numbers, it can inflict considerable losses on apiculturists.

In a few of the higher wasps, certain mouthparts have become modified so as to form tube-like structures up to 1 centimetre ($\frac{3}{8}$ in) long, through which nectar can be sucked. *Polistes* even stores nectar for the brood in addition to animal food, and acts as a regular pollinator. Yellowjackets (*Vespa* or *Vespula*), although carnivorous, can, by sheer force of numbers,

There are relatively few known cases where ants play a direct role in pollination, but *Orthocarpus pusillus* relies almost exclusively on these insects and has thus become known as the 'ant-plant'. It is a low-growing herb of the figwort family, with non-showy flowers. Flying insects would not be satisfied with the small nectar reward offered by its flowers, though it is adequate for the tiny ant.

The European figwort (*Scrophularia nodosa*) is thought to be pollinated exclusively by wasps which are attracted by the scent and flesh-coloured markings of its flowers. While collecting a nectar reward, pollen is deposited on the underside of the wasp's head.

Flowers of the European helleborine, *Epipactis*, a woodland species, are pollinated by *Vespula* wasps. Such a visitor is rewarded with a copious supply of liquid nectar, and, while drinking, clubbed pollinia will become attached to its forehead, mouthparts or even eyes. The wasp would normally wipe these unwanted objects off with its feet, but the nectar which it has been drinking is in fact mildly toxic and makes the unsuspecting insect 'drunk'. Weakened by the effects, the wasp is unable to clean itself or indeed to fly, and can only stagger from flower to flower transferring pollinia as it goes.

become very beneficial as pollinators in late summer, once their larvae have been raised. These animals have good colour vision and an excellent time sense, and can also be trained to certain odours. Such characteristics certainly tend to make them good pollinators. The question of whether or not there are special wasp-flowers has been a hotly debated issue ever since the famous botanist Hermann Müller, in 1873, characterized figwort (*Scrophularia nodosa*) as one. In the Netherlands, certain terrestrial orchids (*Epipactis* spp) are referred to as 'wasp orchids'. Their flowers present, on the labellum, glistening nectar, which is eagerly lapped up by yellowjackets, with resulting pollination. Another yellowjacket favourite is snowberry (*Symphoricarpos*).

Figs and fig wasps: an ancient partnership

Economically and spiritually, fig trees – members of the large genus *Ficus*, in the mulberry family – have played an important role in several cultures. Among others we might mention the classical Mediterranean fig (*Ficus carica*), which is now grown in subtropical areas all over the world; the sacred banyan tree (*Ficus bengalensis*) of India, under which the Buddha and his disciples used to meditate; and the sycamore fig (*Ficus sycomorus*), which for thousands of years has been grown in Egypt as a shade tree, and which also produces valuable wood and tasty fruits. Although in that country sycamore figs have from time immemorial been propagated only in a vegetative manner, they do reproduce sexually in Kenya and Yemen, where they probably originated. As the Israeli botanist Joseph Galil and his students have demonstrated, they present us with a pollination story every bit as fabulous as a tale from *The Arabian Nights*. Indeed, the relationship between fig trees and the tiny gall wasps that pollinate them represents one of the most magnificent cases of symbiosis known to science.

Before we can explain this, however, we must first answer the question: what exactly is the fig we eat, and what is a gall wasp? Essentially, a fig represents a complete inflorescence which in the course of evolution has become urn-shaped and fleshy. It looks like a hollow pear with its blunt side up and its inner surface covered with tiny male and female flowers. The rounded broad tip has a small opening, the eye or ostiolum, surrounded and also partly plugged by a large number of scales, which may overlap. In most fig species, the flowers placed close to the ostiolum are male, the ones deeper down female. As a fig ripens, the 'pear' becomes more and more soft, juicy and sweet. The hard, tiny grains which we notice when we chew on a ripe fig are the stony kernels or seeds of the individual fruits or fruitlets, scattered on the inner surface of the beaker and each preceded by a female flower.

Gall wasps are insects whose females, in general, deposit their eggs in the leaves, stems, flowers or roots of plants. The plant cells respond by forming a mass of tissue which ultimately makes up the main body of a special structure, the gall, which is extremely characteristic – in shape, size, colour and texture – of the particular insect/plant combination we are dealing with. Often, the shape is bizarre. On roses, for instance, we may find the famous moss- or bezoar-galls, which ultimately resemble pieces of very delicate blood coral, while oak leaves sometimes display miniature red-cheeked 'apples'. As the young gall-wasp larva that hatched from the egg grows, so does the gall. There must be a continuous interaction between the young insect and the tissues that surround it, and it is obvious that in many cases the natural hormone balance in the plant cells is completely upset. In figs, an individual female flower (not much more than a pistil, really) can be transformed into a gall after the female wasp has deposited an egg into it by pushing her ovipositor all the way down through the length of the style, from the stigma to the ovary. The mature gall (not very fancy, in this case) resembles a tiny bottle with a narrow neck and a round bottom part which contains the pupa. The walls of this little prison, at that time, are dry, woody, paper thin and brittle.

In the humid forests near Nairobi, Kenya, *Ficus sycomorus* trees keep producing fruits all through the year. Each requires six to seven weeks to reach complete maturity. Wind pollination can be ruled out because the scales in the ostiolum are densely packed; and self-pollination is ruled out because the female flowers reach maturity some four weeks before the male ones do. When the anthers of the latter liberate their pollen, the female flowers are totally spent already. Parthenogenesis, that is the formation of viable seeds without pollination and fertilization, does not play a role either. So, who pollinates these figs?

A clue to the situation is provided by the fact that close to the ostiolum of a ripe fig one may find one or several round holes, about 1.5 millimetres ($\frac{1}{16}$ in) in diameter. These are the open ends of small tunnels running through the fleshy wall and connecting the interior cavity of the fig with the outside world. These tunnels must have been made by some agent. If a fig is opened a few days before complete maturity, one usually finds inside a motley crowd (sometimes comprising several hundred individuals) representing six species of small gall wasps, winged and wingless, blind or with conspicuous eyes, and ranging in colour from light brown to almost black or metallic green. The most important of the wasps found in *Ficus sycomorus* figs is *Ceratosolen arabicus*. The females of this species – long-winged, light-brown animals about 2 millimetres ($\frac{3}{32}$ in) long – are the animals one can see emerging from the holes we just mentioned, usually towards the end of the night or in the wee small hours of the morning. They fly off to neighbouring fig trees bearing inflorescences that are in the female (young) stage. One mature fig can yield considerable numbers of such females. The attraction which the young inflorescences exert upon them must be very powerful, because reaching their interior cavity (which is what the female insects try to do) requires considerable strength, effort and perseverance. Even though the animals are beautifully adapted to the task of fighting their way through the ostiolar scales, it is a fact that many lose their wings, and often even parts of their antennae, during the passage. Still, at least *some* females (usually 10 to 20) overcome all obstacles and end up in the interior of the fig, which at this stage is only 12 to 15 millimetres ($\frac{1}{2}$ to $\frac{5}{8}$ in) long while its wall is 3 millimetres ($\frac{1}{8}$ in) thick. Once inside, they proceed immediately towards the female flowers, of which there are hundreds standing close together on the inner wall of the fig with their stigmatal hairs woven together to form a felty mat, the synstigma. The female flowers, which at this stage are receptive, are of two types: long-styled ones, with a style of about 1.5 millimetres ($\frac{1}{16}$ in), and short-styled flowers, with one of about 0.8 millimetre. The long-styled flowers are sessile while the short-styled ones rest on pedestals, with

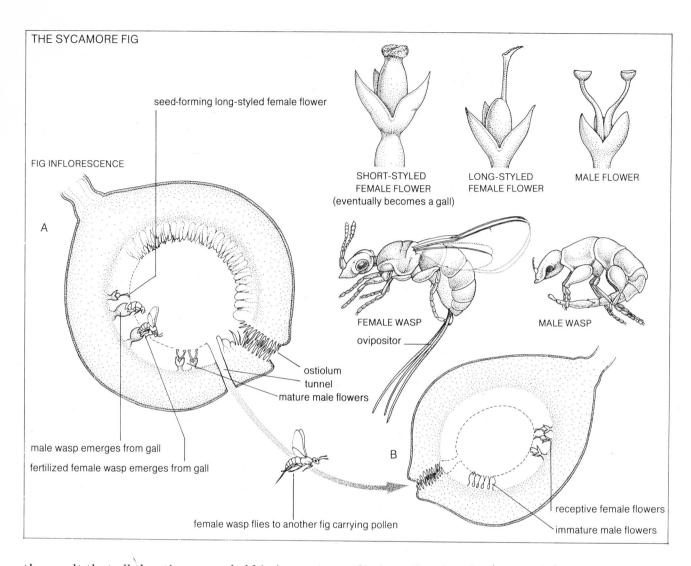

THE SYCAMORE FIG

FIG INFLORESCENCE

seed-forming long-styled female flower

A

SHORT-STYLED
FEMALE FLOWER
(eventually becomes a gall)

LONG-STYLED
FEMALE FLOWER

MALE FLOWER

FEMALE WASP

ovipositor

MALE WASP

ostiolum
tunnel
mature male flowers

male wasp emerges from gall

fertilized female wasp emerges from gall

B

female wasp flies to another fig carrying pollen

receptive female flowers

immature male flowers

the result that all the stigmas are held in in one (curved) plane, forming the synstigma. The ovipositor of *Ceratosolen arabicus* has exactly the right length (0.8 to 0.9 millimetre) to make it possible for a female wasp to deposit an egg in the ovary of a short-styled flower. The act requires about a minute. A female sticking her ovipositor into the style of a long-styled flower perceives her mistake after about five seconds and retracts the organ without depositing an egg. Although pollen is deposited on all the flowers, only the long-styled ones will produce seed: the short-styled flowers develop into small galls and produce only a new generation of wasps. Within a short time, the females that have penetrated into a young fig can deposit many eggs. But where are (or were) the males that were necessary to produce these viable eggs? Inspection of the females emerging from a mature fig reveals that they always contain sperm in their seminal bladder. So they must have been fertilized *before* leaving the mature fig. To fully understand this situation, we must first follow the development of a gall flower with its wasp-larva inside all the way – a process which in nature requires about two to three weeks. Not more than a few days after the larvae have pupated, a number of young wasps of the new generation hatch. With their strong mouthparts, they bite a hole in the wall of the woody gall-flower that held

Inflorescences of the sycamore fig (*Ficus sycomorus*) are almost entirely dependent on the complete life-cycle of the fig wasp (*Ceratosolen arabicus*) for their pollination. A nearly ripe fig (A) contains developing seeds and fig wasps. There are several distinct stages – male wasps emerge from their galls, fertilize the female wasps still within their galls and then bore a tunnel to the exterior through which pollen-carrying females will subsequently escape. Female wasps fly to a young fig (B), force their way through its ostiolum and, once inside, lay their eggs in the short-styled flowers and fertilize the long-styled flowers.

them prisoner, and emerge. These animals look totally different from the winged females with whom we are already familiar. They are small, wingless, practically blind, and clumsy, and possess a relatively big, strong head with heavy mouthparts. Their abdomen is slender and protracted into a thin tube, which is often kept in a forward position between the legs like the tail of a frightened dog. These animals are the males of *Ceratosolen*. Clambering around among the gall-flowers, they will inspect these one by one, obviously in search of females. Having found a suitable gall-flower, the male will gnaw a little hole in the woody wall just large enough to insert the tip of his abdomen and fertilize the young female inside. It is not until several hours later that these females emerge, having enlarged the holes in their gall-flowers. *Ceratosolen arabicus* is clearly protandrous, with the males hatching about a day before the females. Later on, we shall discuss the remarkable timing mechanism involved here.

Once fertilization of the females has been achieved, the task ordained for the wingless males of *Ceratosolen* is still not over. Up to 30 of them may assemble in the upper part of the fig, near the ostiolum, and in a combined effort these animals will gnaw a tunnel through the tough wall which is now almost 5 millimetres ($\frac{3}{16}$ in) thick. Progress is slow, and an unfinished tunnel may be filled up almost totally by the males doing their laborious job. After the breakthrough, the males, which are very sensitive to light, retreat to the fig's interior, where they soon die. Their whole life is spent within the fig and it is therefore not totally surprising that they should lack wings and functional eyes. The females emerge by crawling through the tunnels made by their doomed mates. Since their own mouthparts are not too strong, they could never do the tunnelling job themselves, to say nothing of the fact that they would almost certainly damage their wings in the process. Were that to happen the females would be unable to fly to young figs on other trees to deposit their eggs; and the cycle could not be repeated.

Pollen transport in *Ficus sycomorus* and some other fig species is unique. The moment at which the male flowers begin to shed their pollen coincides neatly with the completion of the tunnel-digging job of the male *Ceratosolen* wasps. It is logical to assume that the females pick up the pollen before exiting through the tunnel to fly to an immature fig, where the female flowers are available and receptive while the male ones are still closed. For the pollen transport, each female has on the underside of her thorax two small, elongate pollen pouches which can hold between 2,000 and 3,000 pollen grains each. Pollen-loading, Galil found, is a very deliberate and lightning-fast act on the part of the gall wasps, which may be carried out some 40 times in succession, with different male flowers as the pollen-source. The unloading of the pollen grains in young figs is also a very 'deliberate', positive act. It amounts to a reversal of the loading process, and starts while the ovipositor is still far down in the style of a flower. The pollen goes from the thorax-pouches to the leg-baskets and from there, in a series of quick movements, to the stigmatic hairs, which form the felty mat (synstigma) over the female flowers.

The fig/wasp relationship: a subtle equilibrium

From the point of view of the fig plant, *Ceratosolen* – and other wasps with a similar life-style – are both 'good' and 'bad'. By converting certain fig flowers into galls, these wasps destroy many ovules and thus depress seed production; in addition, they force the plant to funnel some of its resources into the developing galls, and this may be a more 'expensive' proposition than normal seed development. The other side of the coin is, however, pollination: without the gall wasps, fig plants simply would not reproduce from one generation to the next. This is clearly shown by the fact that in Egypt, Israel and Lebanon, where for some reason *Ceratosolen* is lacking, *Ficus sycomorus* can be propagated only by cuttings. What we are confronted with is a fascinating cost/benefit relationship. In the magnificent reproductive symbiosis which figs and their wasps represent, neither of the two partners can be allowed to start dominating the other. If, for instance, all the female flowers within a fig were short-styled, they would all be converted into galls, and there would be no seeds and no new generation of fig trees. Indeed, some investigators have interpreted the evolutionary development of long-styled flowers by figs as some sort of defence reaction, preventing hegemony of intruding gall wasps (no matter how temporary and ultimately suicidal such a hegemony would be). If the figs die out, then the gall wasps are doomed also.

Pollination by moths

Hawkmoths, also known as sphinx moths (Sphingidae) because of the peculiar position adopted by their caterpillars when disturbed, are the nocturnal counterparts of hummingbirds. Like them, they normally feed on the wing, and, since they are usually quite large, fly at high speeds, and operate essentially as warm-blooded animals, their energy requirements and their consumption of nectar as 'fuel' are high. The famous death's head hawkmoth of Europe, which in the past was associated with war and pestilence, is known to raid beehives, at least occasionally. It may consume a good teaspoonful of honey at a single sitting. Hawkmoths are highly specialized flower visitors, equipped with a long, thin and very flexible proboscis that is kept coiled up when the animal is not feeding but which can be stretched out to take nectar. The tongue is usually extended just as the moth reaches its floral target. The moth aims for the nectar guides with superb marksmanship: indeed the scientist Fritz Knoll chose a small hawkmoth species to demonstrate, in a most elegant fashion, that nectar guides do actually guide. Keeping his animals in a large cage, he fed them sugar water or honey from artificial flowers so that their tongues became quite sticky. He then mounted flowers of butter-and-eggs (*Linaria vulgaris*), characterized by very distinct orange nectar guides, behind a vertical plate of colourless glass. The hawkmoths, trying to stick their tongues into the flowers, were of course unable to reach them, but left sticky marks on the glass corresponding exactly with the nectar guides. Subsequent spraying of the plate with red-lead powder, followed by heating, produced a very convincing permanent record. Hawkmoths have good colour vision, essentially similar to that of bees, and like them they respond to the near ultraviolet. Their sense of smell is also highly developed.

Not surprisingly, flowers that cater to hawkmoths open in the evening as the result of the activity of a biological clock, and display their overwhelming fragrance at that time. They are snow-white or light-coloured, offer no

Six-spot Burnet moths (*Zygaena filipendulae*) are active by day and are frequently found in large numbers among meadow flowers in Europe. Here several of these brightly coloured insects forage among the flower heads of the water mint (*Mentha aquatica*), a species which is characterized by its strong aromatic scent and which is also frequented by bees and numerous other flying insects.

landing platform and may have fringed petals – possibly for guidance. Many have both visual and olfactory nectar-guides. The corolla tube is long and narrow, something which discriminates against other, shorter-tongued, visitors, and there is an abundance of nectar, which often is present in a hollow spur. In the Madagascar orchid *Angraecum sesquipedale* that spur is exceedingly long. Sesquipedale means 'a foot and a half', and although very few spurs of this orchid actually reach that length, it seems obvious that the natural pollinator has to be very long-tongued. Long before it was found in nature, both Charles Darwin and Alfred R. Wallace – the fathers of the evolution theory – predicted that the pollinating animal would be a hawk-moth. They were vindicated some 40 years after making their statement. The moth is a variety of the African species *Xanthopan morgani* and, under-standably, it has received the name *forma praedicta*.

Typical hawkmoth flowers are evening primrose, four o'clock flowers, jimson weed, stephanotis, certain tobacco flowers that are white and very fragrant, and most honeysuckle species. Also, *Aquilegia pubescens*, a pale-flowered columbine found in the Rocky Mountains at elevations up to 2,800 metres (9,200 ft) is favoured by hawkmoths. At night, this montane environ-ment may be too cold even for the versatile hawkmoth and it is suspected that the moths there may switch their activities to the daylight hours.

Other important pollinating moths are the various species of *Plusia* (silver-Y moths), which sometimes occur in countless numbers. The small yucca moths (*Tegeticula maculata*) are totally dependent on yucca flowers for their survival. Their females display a type of highly specialized be-haviour which at first sight does not seem to make any sense at all. Having

Stately spires of *Yucca whipplei* tower above the Californian landscape (*above left*). Their pendent flowers provide a single species of moth, the yucca moth, with a breeding ground and food for its young larvae. The female moth (*above right*) first pollinates a flower with a ball of sticky pollen, then lays a number of eggs in the flower's ovary. Of the seeds that develop, some are devoured by the larvae of the moth, while the rest reach maturity as ripe seeds.

74

landed on a flower, a female prepares a ball of yucca pollen – the only task for which her front legs and mouthparts are suited. She then stuffs the ball into the hollow stigma. With pollination (to be followed by fertilization) thus secured, the ovules begin to develop into seeds. The pollinating female also deposits eggs in the lower part of the pistil in a separate 'deliberate' act, and the larvae that hatch from these feed on the young seeds. The moths can therefore easily be seen as parasites. However, they are at the same time beneficial: enough fertilized ovules are usually spared to secure the production of some viable seeds, and if it had not been for the moth's activities, there would not have been any developing seeds in the first place.

Pollination by butterflies

The great difference between moths and butterflies, which together form the group Lepidoptera, is that most butterflies are sun-lovers that like to perch while feeding. They too have a long and slender proboscis, a good colour sense (which for the whites and the swallow-tails includes red) and an excellent sense of smell. With few exceptions, they are nectar-feeders also. From these characteristics we can easily deduce the characteristics of typical butterfly-pollinated flowers. They are open in the daytime, produce a goodly amount of nectar, possess a long, thin corolla tube (often with a nectar-spur) and are generally vividly coloured (often red) or white. They also provide their butterfly visitors with a platform to land and walk on, acting either as an individual unit (as we see in some *Impatiens* species and

Butterflies tend to prefer inflorescences which provide them with a relatively flat platform on which to stand while probing for nectar. The bright flowers of *Lantana* are borne in just such an arrangement, and here attract an *Agraulis vanillae* butterfly.

in *Bougainvillea*) or, if they are small, in combination with many other flowers in the same inflorescence. The flat-topped inflorescences of verbena, lantana, red valerian, milkweeds and various composites provide excellent examples, as do the gently rounded ones of butterfly bush (*Buddleia*) with their heavenly fragrance. Purple loosestrife (*Lythrum salicaria*) and some violets (*Viola*) represent yet another type of butterfly-pollinated flower.

Pollination by birds

In Europe, sparrows often tear crocus flowers apart to get at the nectar. However, this is just an incidental relationship. Highly specialized flower-birds would need nectar on a year-round basis, and the flowering season in Europe is so short that this condition cannot be fulfilled. Unfavourable geographical circumstances have, furthermore, made it impossible for bird pollination to take hold in Europe by immigration of flower birds from the south in summer. In America, such immigration is easier; some humming-birds even make it all the way to Alaska. In California, a few hummingbird species are permanent residents, while in Seattle in the State of Washington a most gratifying development of the last ten years has been the establishment – on a year-round basis – of a population of Anna's hummingbird. This was possible only because so many people there put hummingbird feeders in their gardens, filled with dilute honey or sugar solutions.

In the tropics and the southern temperate zones, bird pollination is at least as important as insect pollination. Roughly one-third of the 300 families of flowering plants have at least some members with bird-pollinated flowers. Conversely, there are at least 2,000 species of birds in 50 families that feed regularly or occasionally on nectar, flower-inhabiting spiders and insects or

Right The tongue of the yellow-plumed honeyeater (*Meliphago ornata*) is brush-like at its tip and so an ideal implement for lapping up nectar and pollen. Such birds are the main pollinators of many Australian plants in the myrtle and protea families; in this case *Eucalyptus preissiana*, the bell-fruited mallee.

Hummingbirds, such as these two Californian species, are capable of very controlled hovering flight, enabling them to feed from nectaries within suitably oriented flowers without the need to land on them. *Below left* – a female rufous hummingbird (*Selasphorus rufus*) visits the flowers of scarlet gilia (*Ipomopsis aggregata*). *Below right* – a male black-chinned hummingbird (*Archilochus alexandri*) seeks nectar within the long floral spurs of *Aquilegia formosa*, a columbine found growing along mountain streams.

(rarely) pollen. Even some fruit-eaters among the tropical birds will occasionally consume nectar or the special, solid food-tissue offered by certain flower species. In some parts of Central America, the proportion of bird-pollinated flowers is as high as 50 per cent. In New Zealand, which was originally without bees, we find a fairly high number. A real paradise for flower-birds, however, is Australia, where more than 100 bird species are involved with more than 1,000 different species of flower. In southwestern Australia alone – which forms a special biological province – there are 560 plant species in sixteen families that are ornithophilous, that is, bird-pollinated. Bird pollination in this part of the world must be very old indeed, for in some plant genera found only in Australia we find that all the species are bird-pollinated.

Those flower-birds that are the most specialized, as evidenced by the shape of their bills and by the structure of their tongues (which are often tube-shaped or provided with brush-like tips), belong to only eight families: the hummingbirds, sunbirds, honey-eaters, brush-tongued parrots, white-eyes, flower-peckers, honey-creepers (or sugar-birds) and Hawaiian honey-creepers. Determination of their food-plants is often possible by looking at the pollen attached to their bills and feathers. This has been done even with museum skins, and even in some cases with species that are now extinct.

Most birds are active in the daytime and rely on their keen vision to find their targets. Being warm-blooded, they have a high metabolic rate, and especially when they are small (as most hummingbirds are) their caloric intake per gramme of body-weight is large. This is what makes humming-birds such magnificently efficient pollinators. Although few people who watch these jewel-like creatures flit from flower to flower in bright sunlight are aware of it, hummingbirds are constantly on the verge of death: they *must* visit hundreds, nay thousands, of nectar-rich flowers every day, simply in order to stay alive. The darkness of night, which makes feeding impossible for them, poses a grave problem. However, the smaller hummingbirds have solved it very elegantly by going into hibernation every night. Their temperature drops, and their rate of metabolism may go down to about one-fifteenth of the peak daytime value, so that they can save a great deal of energy. In the wee small hours of the morning, under the influence of a biological clock, their metabolism and temperature begin to go up again, and when daybreak comes they are ready to tackle another day.

As mentioned earlier, birds have excellent colour vision and appear to favour red. Very recently, it has been demonstrated that at least some hummingbirds can perceive the near ultraviolet as a colour as well. In the realm of vision, this makes hummingbirds the best-endowed pollinators of all. In contrast, the birds' sense of smell is very poorly developed. All the above bird characteristics – positive and negative – are reflected in the flowers that cater to them. For one thing, they have no odour. In this context it is very illuminating indeed to compare the odourless, orange-coloured flowers of the honeysuckle *Lonicera ciliosa* of the Pacific Northwest, which is pollinated by hummingbirds in the daytime, with those of most other honey-suckles, which are light-coloured, proverbially fragrant and nocturnal, and are pollinated by hawkmoths. The amount of nectar produced by bird-flowers can be prodigious; a thimbleful in the case of certain coral-trees (*Erythrina*), a liqueur-glassful in that of the spear-lily (*Doryanthes*). Small

The vivid red flowers of *Passiflora vitifolia* are borne on bare stems close to ground level within the tropical jungles of Costa Rica. Surprisingly, perhaps, this is actually a climbing vine and bears its leaves high up in the forest canopy, but for pollination it relies on hermit hummingbirds, which specialize in feeding at lower levels. Red is a colour characteristically favoured by hummingbirds.

wonder that Australia, with its abundance of bird-pollinated flowers, is now a paradise also for the very versatile domestic honey-bees that have been introduced there.

The colours of bird-pollinated flowers vary enormously. In general, people visiting tropical gardens in the daytime are struck by the dominance of red and the absence of fragrance there. In Hawaii, however, the flowers visited by birds are (or were) usually white. Fierce parrot colours, with complementary colours appearing side by side, are striking in the pineapple family, or Bromeliaceae, where an inflorescence may be green plus red plus yellow plus blue; in *Aloe*, where we see yellow or red plus green; and in the well-known bird-of-paradise flower, *Strelitzia* (which is *not* pollinated by birds-of-paradise), which is orange-yellow plus blue. The question why red is so common in bird-flowers, at least in North America, is still controversial. Many – perhaps most – floral biologists see a direct connection here with the preference for reds which many birds display. However, Verne and Karen Grant, who have written a magnificent book on hummingbirds and their flowers, have challenged this explanation, claiming that in the tropics, with a more stable population of flower-birds than the US, red is not all that predominant. They also point out that the flowers of *Marcgravia* species with their striking nectar-tubes, usually thought to be bird-pollinated, are not red. In North America colours other than red, the Grants feel, have been 'pre-empted' as attractants by flowers that are insect-pollinated; this left only red as a characteristic that could be utilized by flowers producing large amounts of nectar (for hummingbirds) while at the same time excluding   79 bees. Such an arrangement, according to the Grants, is especially important for hummingbirds because, as migratory animals, they are forced to operate in a number of different environments where quick recognition of their food-source is essential. A legitimate question to ask, however, is: why do birds display a definite preference for red berries? And why is it that, in the plant family of the Gesneriaceae, the South American members (which cater to birds) are characterized by special aurone pigments that provide fierce colours, while Asiatic Gesneriaceae (which for the most part are not ornitho-philous) contain the 'normal' anthocyanin pigments? This situation certainly provides a strong argument for *special* colour adaptations of flowers to hummingbirds. As to the *Marcgravias*, many biologists have come to the conclusion that they are bat-flowers rather than bird-flowers.

Ornithophilous flowers are open in the daytime. Their size and shape vary widely from species to species and if they have a flower-tube it is wider than that found in butterfly- or moth-pollinated flowers. They are sturdily con-structed as a protection against the probing bills of their visitors, and their vulnerable ovules are kept out of harm's way in a so-called inferior ovary (one that is placed beneath the floral chamber, as we see in *Fuchsia*), or, con-versely, in an ovary that sits on a special stalk, as it does in members of the caper family (Capparidaceae). In some species, the ovary is hidden behind a screen formed by the fused bases of the stamens. Usually, there are quite a few of these; they often stick out of the flower, in which case they are brightly coloured, and in general they are so strong as to remind one of wire. The visiting bird is normally touched on the head or breast as it feeds. How-ever, a number of ornithophilous flowers, for example some species of *Hyptis* and of *Loranthus*, a tropical mistletoe, are explosive and cover their visitors thoroughly with their pollen. In other cases, for example in *Strelitzia*, the pollen grains are connected by sticky threads. Consequently, the pollinator will always take a large number of them to the stigma of the next flower visited, and it is not surprising that in cases like that the number of seeds per fruit is large.

Hummingbirds, which are found only in the Americas, normally feed on the wing. The corresponding feature of American bird-flowers such as fuchsias and salmonberry flowers is that they hang down or are downward-facing. They also lack the landing-platform (the lip) found in so many bee-flowers. In Asia and Africa, on the other hand, flower-birds – with very few exceptions, such as the *Arachnothera* of Malaysia – do not hover, and accordingly the plants they pollinate offer them a landing-platform or pro-vide perches in the form of small twigs near the flowers. It is interesting that some South African and Australian plants that cater to birds flower close to the ground. A good example is the Australian *Anigozanthos humilis*. It is 'served' by wattlebirds, which have the habit of drinking while standing up and prefer to hop from plant to plant.

Pollination by non-flying mammals

It is quite possible that in the Upper Cretaceous period, before bats appeared on the scene, small climbing marsupials and (later) even prosimians feasted upon flowers. These must have adapted to the animals, so that gradually a special mammal syndrome developed. In most parts of the tropics, flowers

The Australian honey possum (*Tarsipes spenserae*) is adapted to a diet of nectar, pollen and small insects. It has a few small, peg-like teeth, a long tongue, grasping feet and a prehensile tail. It is exclusively arboreal and frequents banksias and eucalypts such as this coral gum.

that possessed it were later taken over by the competitive and well-equipped bats. However, in certain isolated parts of the world such as Australia and Madagascar, characterized by an ancient flora and fauna, we still find remnants of the original situation. (We must add, though, that even in those areas more modern types, such as rats and mice, have also become involved.)

In various parts of Australia, *Banksia* and, in South Africa, *Protea* species are regularly visited for nectar not only by mouse-like nocturnal marsupials (pygmy possums and honey-possums, of which *Tarsipes* is the most specialized) but also by native rats (*Rattus fuscipes*). All in all, no fewer than 21 species of Australian marsupials may enter into relationships with

flowers. In the Cape Province of South Africa, scientists found that low-growing *Protea* species are exploited by nocturnal wild mice of the genera *Acomys*, *Aethomys*, *Praomys* and *Rhabdomys*, and also by shrews (*Crocidura*). There is a beautiful correspondence between the shape of the flowers, which are formed very close to the ground, and the snouts of the animals. The dull-reddish flowers, which at night develop a yeasty odour, are combined in dense heads, and the nectar, produced in copious amounts, collects in the centre of these. In Madagascar, mouse lemurs (*Microcebus*) are involved in pollination. Recently, in 1980, it was discovered that in the cloud forest of Costa Rica, certain nocturnal mice, notably *Cryzomys* and *Peromyscus*, visit the flowers of a hitherto undescribed species of *Blakea*, an epiphyte in the family of the Melastomataceae, for the sake of the nectar. Significantly, this fluid is secreted only at night. However, the colour of the flower and the absence of sweet fragrance provide proof that there is no hawkmoth pollination. Monkeys, although often destructive, have recently been implicated in pollination in both South America and West Africa.

Pollination by bats

Historically, bats have always been surrounded by superstition and fear. The fact that they are mostly nocturnal has not improved their reputation and so they have been used as ingredients in both witches' brews and horror movies. It is, of course, true that there *are* blood-sucking vampire bats in Central and South America, and that in the USA rabies is sometimes transmitted to humans by bats. However, this should not blind us to the many fascinating features these animals display, such as their ability to locate – and thus avoid or catch – small objects in the dark by sonar. And, undeniably, certain bats play a very positive role in nature as pollinators of tropical plants. Yes – just as in bird pollination, chiropterophily, or bat pollination, is an essentially tropical phenomenon, even though a certain level of bat pollination is maintained in the southwestern USA by temporary immigration of Mexican bats in the summer. Even more so than hummingbirds, certain bats have become pure 'flower-animals' – especially those in the group of the Macroglossinae or 'big-tongues', found in southern Asia and the Pacific and exemplified by the genus *Macroglossus*. Their protein requirement is met entirely by pollen, which they deliberately collect and consume in great quantities, along with nectar. *Glossophaga*, a New World bat, has the same life-style.

Let us first look at the characteristics of these fascinating animals, and then at the corresponding features of the flowers they serve so well. In spite of the saying 'blind as a bat', bats have reasonably good eyesight; however, they may well be colour-blind. In any case, good colour vision would not do them much good in the semi-darkness in which they normally operate. They have a keen sense of smell, displaying a preference for odours which we humans find definitely unpleasant: mouse- or urine-like, stale, musty or rancid, often reminiscent of cabbage, butyric acid and sweaty feet. The sonar sense of flower-pollinating bats, at least in Asia, is by no means as well developed as it is in other bats.

Macroglossus bats are small, sharp-snouted animals with long tongues that can be stuck out very far and have special projections (papillae) plus,

The Namaqua rock mouse (*Aethomys namaquensis*) feeds by night on nectar from certain *Protea* species which bear flowers very low to the ground, generally in a downward-facing position.

83

At rest on a wall after a feast of nectar, this *Leptonycteris* bat's furry snout and head are heavily powdered with the yellow pollen of a century plant (*Agave palmeri*). The inflorescences of this species are almost entirely composed of stamens and pistils; colourful petals would serve no purpose in luring night-time visitors.

in some cases, a soft brush-like tip – devices which enable them to pick up rapidly the copious pollen/nectar soup which bat-flowers offer. Likewise, *Musonycteris harrisonii*, a close relative of *Glossophaga*, has a tongue which, at 76 millimetres (3 in) is almost as long as its body (80 millimetres, $3\frac{1}{4}$ in). *Glossophaga* itself has hairs that are especially adapted for pollen transport. They have minute scales, comparable in size to the scales found on the hairs of a bumblebee's abdomen. Normally, it is a bat's head that becomes dusted with pollen; thus, transfer of the precious powder to the pistils of other flowers is no problem. Considering the extreme food specialization, it is not surprising that teeth (in *Macroglossus*) are almost entirely lacking. The few found in the males – in some cases – are used for fighting and not for eating. During the latter activity, these bats hook themselves into the petals of the flowers (which they probably locate by smell) with their thumb-claws. However, some American flower-bats seem to hover like hummingbirds while obtaining their food.

The typical bat-flower is large, sturdy, bell-shaped with a wide mouth, and either snow-white as in the baobab tree (*Adansonia*) or drab in colour as in the sausage tree (*Kigelia*). The drabness is probably correlated with the lack of colour vision in bats, while the whiteness makes sense because it provides conspicuousness by contrast with the darker background. The strong odour of bat-pollinated flowers agrees with the animals' preferences. Food is offered in abundance. There are two ways in which bat-flowers have increased pollen production during the process of evolution: by increasing the number of stamens (to about 2,000 per flower in baobab, as we have seen) and by making very large anthers, as in *Agave*. The amount of nectar in *Ochroma grandiflora*, a relative of the balsa tree, may be as high as 15 milli-litres ($\frac{1}{2}$ fl oz) per flower. The flowers of a bat-pollinated *Oroxylon* have a neat tipping mechanism that enables them to dole out their nectar in small portions. As a result, about 36 visits are necessary to exhaust a single flower's total supply. Other bat-flowers offer their visitors special food-bodies in the form of succulent petals or sweet-tasting bracts. These flowers are served by fruit-eating bats such as the notorious flying foxes (*Pteropus*) that can inflict such severe damage in apple orchards in Queensland.

Another striking adaptation, superbly demonstrated by the sausage tree, *Kigelia aethiopica*, is that the flowers dangle beneath the crown, suspended by thin, rope-like branches. In other species, they may be placed on the main trunk or the big limbs. This makes the flowers easy to approach or circumnavigate – a definite advantage to flower-bats with their poor sonar-sense. The pagoda shape of kapok trees (*Ceiba*) serves the same purpose. We must add, though, that such features may also be connected with the dispersal of fruits by bats – a phenomenon that may exist even in plant species such as cacao (*Theobroma*), which are pollinated by agents other than bats (e.g. by flies). In Arizona, the giant saguaro cactus (*Carnegiea*) and the century plant (*Agave*) are pollinated by *Leptonycteris* bats, although not exclusively. These specialists come in from Mexico. Cup-and-saucer vine (*Cobaea scandens*), often cultivated in European and North American gardens, is the direct descendant of a bat-pollinated South American plant. The famous *Marcgravia umbellata* of Central America produces nocturnal bat-flowers of nondescript colour, combined in a most remarkable hanging chandelier. The hummingbirds seen on them in the daytime get only the leftovers. (Related genera, however, such as *Norantea*, are ornithophilous, with nectar-beakers that are red and obviously designed to attract birds.) *Markea* resembles *Marcgravia* in that it too is an epiphyte with a hanging inflorescence. Over a period of several months, the latter develops up to 200 flowers, of which one or two open each night. They contain a prodigious amount of nectar, but their corollas last for four hours only. Not surpris-ingly, bats take 'season-tickets' for visits to the flowers of both *Marcgravia* and *Markea*. Since the population density of these plants is low, the animals are forced to follow typical 'trapline' methods for getting their food. On the inflorescences of other plant species, such as *Agave* and *Lafoensia* species, bats often forage in flocks. It has been shown (at least in *Lafoensia*) that a given flower is, nevertheless, visited just once or twice, probably because bats recognize previous landings by scent marks that are left on the flowers. Other examples of bat-pollinated flowers are calabash, candle tree, *Areca* palm and certain bananas of the genus *Musa*. *Musa fehi*, for instance, which is found on the Hawaiian Islands, is such a plant and ethnobotanists have used this knowledge profitably: since bats do not occur in Hawaii, *Musa fehi* cannot possibly be a true native but must have been introduced by man.

CHAPTER FIVE

The 'unacceptable' face of pollination

In previous chapters we have seen that the basis of many floral strategies is a mutually beneficial relationship between flowers and their animal pollinators. The animals visit flowers to gather 'reward substances' (nectar, pollen, wax, oils, fragrances) and while doing so inadvertently transfer pollen from flower to flower, so bringing about fertilization.

Inherent in the relationship is a risk – a risk that the substances offered by the flower to reward the pollinators for their services will be removed – stolen or plundered, if you like – without the act of pollination being performed in return. One could compare a flower to an unattended shop with a notice kindly asking customers to deposit the correct amount in the till. An awful lot is taken on trust. Where there is bounty there will be thieves, and this dictum holds true for the world of flowers. Consequently, most flowers have evolved so as to simultaneously encourage legitimate visits by pollinators *and* prevent pillage by thieves. The protective devices of flowers are not as obvious as those designed to attract pollinators, but are nonetheless built into their size, shape and colour; the position of their nectaries; the timing of opening and closing, and so on. The commonest ploy is to put the reward beyond the reach of all but the intended visitor, for example by locating nectar at the end of a long, hollow spur or at the bottom of a long or narrow corolla, or by holding pollen out of harm's way on the end of long, delicate anther stalks.

Some plants employ ants to guard their floral nectar, luring them on to or near to the flowers with nectar from extra-floral nectaries. Experiments have shown that the presence of ants is an effective deterrent against insects intent on pillage. However, as ants also rank among those species that are apt to steal floral nectar, several plants, such as the common teasel (*Dipsacus fullonum*) and the Nottingham catchfly (*Silene nutans*) have evolved mechanisms for keeping ants out of their flowers. The latter deploys a 'roadblock' in the form of a sticky, impassable secretion that covers the stems. We can see that, depending upon the species of plant and its location, ants can either be a threat or a safeguard to floral nectar.

Despite their protective mechanisms, many flowers suffer at the hands of pillagers. Syrphid flies and beetles enter flowers to take pollen; small bees, flies and ants collect nectar from all but the most inaccessible nectaries, while large bees like the bumblebees and carpenter bees, and nectar-feeding birds such as the South American flower-piercers, make holes in the sides of flowers to gain illegitimate access to nectaries.

Comfrey (*Symphytum officinale*), that conspicuous roadside weed so favoured by medieval herbalists for treating human and animal ailments, is a good example of an oft-pillaged flower. Its bell-shaped flowers hang

The stem leaves of the European common teasel (*Dipsacus fullonum*) are united at their base to form a reservoir for rainwater and dew, rendering the route from ground to floral nectar more or less impassable to would-be pillagers, such as ants.

downwards, designed to be pollinated by long-tongued bumblebees such as *Bombus hortorum*. They are, however, frequently robbed of nectar by the shorter-tongued *Bombus lucorum*, which chops little holes in the base of the corolla tube with its powerful mandibles. Honey-bees (*Apis mellifera*), too short in the tongue to reach the nectaries legitimately, and too weak-jawed to cut their own pillage-holes, quickly learn to take advantage of the small bumblebee's dirty-work. Although the tubular, downward-hanging flowers of comfrey successfully exclude flies, wasps, beetles, butterflies and ants, they must still put up with losses to the wrong kind of bee.

A rather interesting case of burglary was observed in 1971 by scientists working on hummingbird ecology in Mexico. They noticed that up to 80 per cent of the flowers of *Penstemon kunthii*, normally pollinated by blue-throated, rivoli and white-eared hummingbirds, suffered pillage by the cinnamon-bellied flower-piercer. The flower-piercer is a professional nectar thief, and has evolved a special beak for the job. The hooked and serrated upper mandible grips the corolla tube, while the sharp lower mandible makes the hole. In the case of *P. kunthii*, the pillage-holes made by the cinnamon-bellied flower-piercer were also visited by the bumblebees *Bombus pulcher* and *B. trinominatus*. The corolla tube is too long and narrow for most

The long bells of comfrey (*Symphytum officinale*) present this bumblebee (*Bombus pratorum*) with a dilemma – she can sense nectar, but her tongue is too short to reach it. Unfortunately for the flower, the ingenious bee has learned to solve this problem by biting through the side of the corolla to obtain access to the nectar – in so doing, this bee never effects pollination.

88

bumblebees to reach the nectaries legitimately, while the thickness of the corolla tissues prevents all but the largest queen bumbles from piercing their own holes.

However, we have only presented one half of the story, for it is by no means always the case that flowers are the victims and animals the villains.

Pollination by foul means

What if a flower were to swap roles and become the 'criminal', deceiving its visitors by indulging in inaccurate advertising, or even physical maltreatment? Bearing in mind the need for pollinator fidelity, could this ever be a viable strategy from the plant's point of view? Indeed it can. And what is more, the relationships between plants and their pollinators contain every nuance of 'criminality' from false representation, through assault and battery, to blatant murder.

Mimicry: specific flower mimics

Although one might expect evolution to fashion flowers so that they are as distinctive and different from one another as possible – to maximize the chances that a pollinator might remain faithful to one particular species of flower – many flowers appear to us to be very much alike. Examples are buttercups (*Ranunculus*), cinquefoils (*Potentilla*) and rock roses (*Helianthemum chamaecistus*), or in Alpine regions, *Ranunculus alpestris*, *Dryas octopetala* and *Chrysanthemum alpinum*.

Perhaps flower mimicry has the effect of achieving a greater degree of pollinator attention and fidelity to the group as a whole, at the expense of other species of plant that are competing for those same pollinators' services. Any advantage so gained would have to more than offset the losses due to pollen wastage within the group. Although there is a degree of deception being practised here, this type of mimicry is hardly at the expense of the pollinators so long as all the flowers offer rewards.

Another sort of mimicry occurs when one species of plant evolves a flower with a close similarity to the flower of another, more common, species that grows in the same vicinity. Not numerous enough by itself to achieve pollinator fidelity, the mimic trades on the occasional mistake made by the pollinators that are visiting the model. In certain parts of Great Britain one can find a little eyebright (*Euphrasia micrantha*), which has lilac or dark pink flowers very much like those of the ling plants (*Calluna vulgaris*) among which it commonly grows. The two species can be found flowering at the same time of year. It has been suggested that they form a mimicry pair, with the ling as the 'model' and the much rarer eyebright as the 'mimic'. The idea is that the imitator profits from the resemblance. Indeed if it can be shown that pollinating insects do not distinguish between the species, one can immediately agree that the hypothesis is plausible.

This certainly seems to be so in the case of the orchid *Cephalanthera longifolia* and its model *Cistus salvifolius*, studied in Israel. The herb *C. salvifolius* has a white flower with yellow pollen and is pollinated by pollen-gathering *Halictus* bees. The orchid sports the same colours but contains no genuine reward; instead, the lip or labellum carries a cluster of yellow hairs, which the bees mistake for pollen. The orchid also produces a sweet scent and manages to fool sufficient bees to achieve a certain amount of fertilization even in the absence of its model, though in this situation the degree of fruit-set is less.

In Central America, female *Centris* bees gather nutritious oils from flowers of the family Malpighiaceae. Dressler, in his book *The Orchids*, states that the same bees are seen to make unprofitable visits to the flowers of *Oncidium* orchids, which mimic those of the oil-producing Malpighiaceae but which actually contain no reward substance.

Food in the window, but none on the shelves

A slightly different type of pollination by deception occurs when the mimic is copying not one particular model species but fashions itself instead to represent a broad range of attractive features – colour, shape and scent – without actually having on offer any of the reward substances that pollinators expect from such flowers. Many species of orchid fall into this category – in fact the number of nectarless orchid species has been put at 8,000 or more, most of which rely upon deception of one sort or another. Many of these have floral structures that observers take to resemble pollen, either in the form of powdery granules or clusters of hairs which could be taken to mimic stamens.

The orchid *Calopogon pulchellus* is a general flower mimic from the eastern USA and Canada. The lip of its flower is covered by a thick mat of hairs that resemble pollen-laden stamens. When a bee lands on the labellum to gather the 'pollen', the lip falls forward under the bee's weight and pivots downwards like a trap-door. The bee tumbles into the flower and falls against the column, which releases a sticky substance on to the bee's back. The shape of the column is reminiscent of a slide in a children's playground, with ridged margins. These ridges fit neatly round the bee's body and the creature slides downwards. Having passed the stigma, the bee collides with the anther and the orchid's four pollinia pop out and attach themselves to the sticky patch on the insect's back. When the bee visits another *C. pulchellus* bloom, the pollinia will be left on the stigma during the downhill slide.

A well-documented example of pollination by deception is that of the calypso orchid (*Calypso bulbosa*), which grows in coniferous forest leaf-litter in North America, Eurasia and Japan. In early spring the orchid produces a single pinkish flower, marked with reddish-brown spots and stripes, that is conspicuous against the dappled green and brown of the forest floor. Its scent is pleasant, but variable in intensity; the flowers contain no nectar, and the pollen is inaccessible to pollen-seeking foragers because it is bound together into a pair of pollinia on the inner, upper surfaces of the flower. James Ackerman studied the pollination of *C. bulbosa* in northern California and concluded that it is a generalized food-flower mimic. Its victims are naive worker and queen bumblebees which are attracted to the flowers on their initial foraging flights. The strategy works because the orchids tend to be locally numerous; batches of bumblebees hatch out in large numbers during the orchid's flowering period; and the coniferous forest leaf-litter is relatively sparse in those places where competing plants bloom. A final refinement is the variability of colour, patterning and scent of the flower,

CALOPOGON

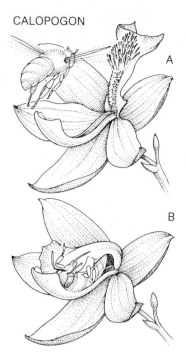

The flowers of the orchid *Calopogon pulchellus* grow with their labellum uppermost; yellow callus growths on the labellum mimic pollen-bearing anthers. When an *Augochlora* bee lands on these (A), its weight causes the hinged labellum to somersault it backwards against the slide-like column (B).

which serves to reduce the ability of a 'wronged' bumblebee to learn to avoid another calypso orchid. The bees do eventually learn to ignore the calypso's deceptive façade, but only after two or three fruitless visits – fruitless for the bees that is, not the orchids.

Carrion, faeces and fungi

So far under this heading of mimicry we have considered only insects seeking foods such as nectar and pollen. But there are many species of insect – bluebottles, flesh flies, fruit flies, certain mosquitoes, midges and beetles for instance – that would not visit flowers for such foods. Their diet consists of faeces, carrion and rotting organic materials of one sort or another. This group of potential pollinators is available to any species of plant that evolves a flower that resembles, in smell or colour or texture, or by a combination of these attributes, the appropriate reward. Many have done so, and the majority of them have opted for flies as their pollinators.

Although the popular conception of an orchid is one of a luscious, fragrant flower, it is among the orchids that one finds the most abundant examples of fly-exploiting flowers. The flowers of these particular orchids rarely contain nectar, or food of any sort, but they display a range of features designed to attract flies over some distance and then to orientate them accurately so that pollen transfer can take place. They include putrescent odours; dull green or purplish colours; spotted patterns or 'warts' that imitate clusters of flies already enjoying a meal; movable parts in the flower to throw or manoeuvre the fly into a position of contact with the pollinia or stigma, and parts of the flower extended to form 'tails' which may help to disperse the scent or, by vibrating in the wind, provide a visual attractant.

Many orchid species of the genus *Bulbophyllum* emit foul odours and are variously specialized to exploit flies of different sizes, from midges to flesh flies. All are characterized by a hinged labellum which tips over when an insect lands on it, throwing the pollinator against the column. In the case of *B. ornatissimum*, large flesh-flies are thrown against the column by the hinged lip, and held there by rows of stiff hooks on each side of the column until eventually they struggle free – carrying the pollinia with them.

Many members of the Pleurothallinidae, a tropical American group, attract flies by means of putrescent odours. Flesh-like colours, hairs and dots act as additional attractants. The greenhoods (*Pterostylis*) from Australia and New Zealand are universally deceitful: they contain little or no nectar, or any alternative reward substance, and like *Bulbophyllum* they have a hinged lip that catapults visiting insects – usually flies, gnats or mosquitoes – into a chamber formed by the petals. The only way of escape is between the lip and the column, where the pollinia and stigma are housed. Once the insect has escaped, the lip returns to its resting position, 'cocked' like the trigger of a gun, ready for the next visitor.

Among the most potent of all 'scent-mimics' is the giant 2.5 metre (8 ft) tall *Amorphophallus titanum* from the jungles of Sumatra. This plant attracts the large *Diamesus* carrion beetles by generating such a powerful and overwhelming smell that men have been known to pass out from taking too close a whiff! The stench bears a close resemblance, to the human nose at least, to a mixture of rotting fish and burnt sugar.

The milkweed family, a group with a propensity for occasionally 'doing-in' its pollinators, includes a South African genus, *Stapelia*, whose flowers resemble rotting flesh in colour and odour. Female blowflies are attracted to the flowers, believing they are carrion. Maggots hatching from eggs laid by them will perish from lack of any suitable food.

Left The large flowers of this lowland species from east Kenya, *Edithcolea grandis*, attract carrion flies. Their colour and scent mimic that of rotting flesh, and the fringe of hairs around each petal shimmers in the breeze, perhaps resembling a swarm of flies.

Below left Another carrion plant, *Hydnora africana*, a root parasite on *Acacia* and *Commiphora* in arid regions of Africa, attracts carrion beetles by scent-mimicry. Large nocturnal flowers appear at the desert surface for just one night, during which they become crammed with beetles.

Right The American eastern skunk cabbage (*Symplocarpus foetidus*) can maintain its inflorescence at a staggering 20–25°C above that of the surrounding soil and air, and emerging early in spring, literally melts its way up through snow and even ice.

Above This *Arum dioscoridis* inflorescence has been coated with liquid crystals (cholesterol-esters) which 'handle' light according to their temperature. They display a colour succession from copper through gold and green to peacock blue as the inflorescence passes through the 25–28°C temperature range.

On a more acceptable size-scale – for something that is producing a foul smell – are various dung-mimicking arums. Lords and ladies (*Arum maculatum*) attracts female dung midges; *Dracunculus vulgaris*, *Sauromatum guttatum* and *Arum dioscoridis* smell like fresh faeces; *Arum orientale* has the sharp tang of spent firecrackers, and *Arisarum proboscideum* resembles fresh woodland fungi. An enormously energetic process is connected with the sudden production of these often overpowering smells. It has been found that in the European lords and ladies, for instance, the spadix, which is laden with starch, may actually warm up by 15°C. The purpose of this heat production seems to be to assist the evaporation of the organic compounds that constitute the foul smell. The performance statistics of this process make interesting reading. At the peak of its metabolic explosion, the spadix of lords and ladies may burn up oxygen as rapidly as a hovering hummingbird – the oft-quoted example of Nature's most energetic machines – consuming oxygen at the rate of $72,000 \, mm^3$ per gram of wet weight per hour. In fact the open-air part of the spadix may 'burn up' over a quarter of its weight in a day.

Flowers as a place to breed

We have seen how flowers achieve pollination by seeming to advertise the presence of carrion without actually delivering it. Because the hungry visitors quickly realize there is no genuine sustenance to be found in these flowers, the pollination mechanisms must be fairly quick-acting if they are to be successful.

93

It is a slightly different matter when the victims of the flowers' deceit are gravid female insects searching for a place to lay their eggs and the males that assemble to mate with them. The insects' behaviour in these circumstances seems to be more determined; less easily put off by a lack of instant success. Moreover, while insects feed *on* things, they tend to lay their eggs *in* things – which involves a good deal of searching, probing, pushing, even digging. What this adds up to from the plant's point of view is less difficulty in encouraging the visitors to become thoroughly involved in the flowers.

Traps and ambushes

Several species of plants have evolved traps, ambushes and detention centres for their unwitting pollinators in order to relieve them of any pollen they might be carrying, and in most cases to send them on their way, hale and hearty, with a fresh load of pollen. The great advantage of physical detention is that each inflorescence or trap needs to attract its visitors only once. Most of these traps share a common plan. There is a chamber formed by various floral parts – petals and sepals – which contains separate male and female floral structures. Often the chamber is an inflorescence, in which case the structures are male and female flowers.

Like all good prisons, the chamber has a narrow entrance equipped with a turnstile or other device that allows one-way passage only. Insects are attracted to the flower, or inflorescence, by the appropriate scent, and by the colour and shape of the structures surrounding the entrance. By various devious means – slippery slopes, pivoting hairs or by their own searching behaviour – the visitors end up in the chamber. Usually only the female structures are receptive at this stage, and they are fertilized by pollen transferred to them from the bodies of the prisoners as they blunder about in an effort to escape. The victims' creature comforts are provided by just sufficient food and moisture to keep them alive, while the flower, or inflorescence, moves on to a pollen-production stage. The anthers mature; the victims become coated with fresh pollen, and are finally released into the fresh air once more. As each trap is a one-shot affair there is obviously disaster in store for the first trap flower to bloom, for there is no chance that its victims will bring in pollen to fertilize its ovules.

Trap-flowers are found in all sorts of environments, and at various heights above the ground. The common lords and ladies (*Arum maculatum*) is found in woodlands and hedgerows, while most pipe-vines are climbers, with some species forming their flowers high up in the canopy. The flowers of *Hydnora africana*, a root parasite, and the inflorescence of *Stylochiton*, an African member of the arum family, form underground traps in the hot desert sand. In contrast, the tubular traps of *Cryptocoryne ciliata*, an aquatic arum lily from Malaysia, are formed underwater. Small fruit flies liberated in a room containing an aquarium with a flowering *Cryptocoryne* plant in it can later be found in the air-filled traps.

The Dutchman's pipe (*Aristolochia clematitis*) – known in Holland as the 'German's pipe' – is found in Europe. As the name implies the flower is a gently curved tube, initially held vertically, with the upper part flanged to form a flag around the entrance, and the lower part inflated to form a chamber in which the pistil and anthers are housed. When newly opened,

A good example of a floral chamber for the detention of small pollinating creatures is found in the arum *Helicodiceros muscivorus* (here shown in cutaway section). A crown-like ring of sterile rudimentary flowers separates zones of male flowers (*top*) and female flowers (*bottom*).

Small gnats, attracted to the flowers of Dutchman's pipe (*Aristolochia clematitis*) by their smell, lose their foothold on the waxy flag and fall down through the vertical, hair-lined tube (A) into the floral trap, which contains both male and female floral parts. If the captives are carrying pollen from another plant of the same species they will pollinate its receptive stigmas in their efforts to escape. The gnats are held within the chamber for two to three days while the stamens mature, dusting them with pollen, then the guard-hairs wither allowing them to escape (B) to repeat the process in another flower.

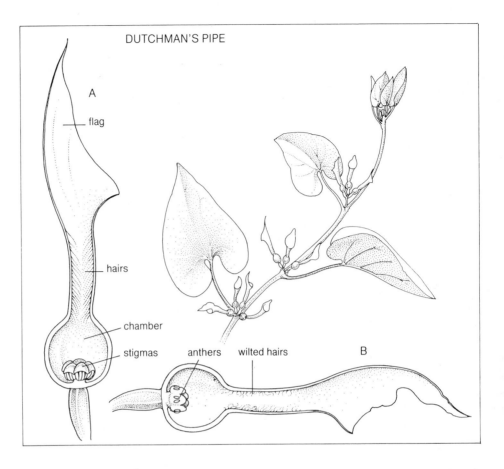

DUTCHMAN'S PIPE

flag

hairs

chamber

stigmas   anthers   wilted hairs

A

B

the flower emits a peculiar smell that attracts small gnats and chironomid flies, especially *Ceratopogon*. The gnats attempt a landing on the vertical surface of the flag, and although most vertical surfaces, or even overhangs, present no problems to these creatures with their hooked and suckered feet, the Dutchman's pipe has a surprise in store for them. The flag is covered with thousands of tiny wax granules and to the gnats it must be like trying to gain a foothold on a hill made of ball bearings. Soon after landing, they find themselves rollercoasting down the tube into the chamber.

In the tube, the gnats pass a battery of downward-pointing hairs, constructed so that they can pivot freely in one direction only – downwards – allowing the gnats to tumble past. If a gnat tries to push its way back past the hairs, they pivot back to the horizontal (at right angles to the wall of the tube) and then lock in position, like the grid over a dungeon. The gnats are securely detained in the dark basal chamber, where the anthers and stigmas are housed. A window in the wall of the chamber lets in light which guides the pollen-bearing captives to the six receptive stigmas. The stamens do not mature for another two to three days, during which time the gnats are maintained on a diet of fresh air and nectar, while pinpoint-size pores allow water vapour to diffuse into the chamber from the turgid cell walls of the chamber, keeping the gnats at the correct humidity.

Finally, the anthers open, showering the gnats with pollen, while at the same time the flower tilts to the horizontal, guard hairs and the flag wilt, and the pollen-coated gnats leave. Despite their ordeal some of these gnats will be drawn to other pipe-vine flowers, effecting cross-pollination.

The Sardinian kidnappers

Long before the inhabitants of Sardinia gained a reputation for kidnapping, a local arum lily had refined the art to perfection. The species concerned, *Helicodiceros muscivorus*, like most good kidnappers, is somewhat rare and selective in its habits. Today it is found only in Sardinia, Corsica and the Balearic Islands. Although we have studied cultivated specimens in Seattle, Washington, and in Oxford, England, our only experience of *Helicodiceros* in the wild was on a tiny rocky island off the southern tip of Sardinia.

The island, Calvoli, is a tangle of shrubs and wild flowers, and hosts a small colony of gulls at its exposed western end. The gulls congregate in the spring to breed, building their nests among the sparse vegetation and boulders that slope down to the sea's edge. Never the most houseproud of birds, the gulls incorporate into their nests, and accumulate around them, a variety of objects including bones, faeces, regurgitated seafood, even dead chicks and addled eggs – a veritable paradise for blowflies. The blowflies recycle this wealth of organic matter, and what is good news for the gulls is good news also for *Helicodiceros muscivorus*, which grows among the gulls'

The huge spathe of *Helicodiceros muscivorus* mimics the coloration of rotting flesh, and the flower emits a stench even stronger than that of the nearby dead gull chicks – a combination of attractants which blowflies find irresistible.

nests, in the fissures and cracks between the boulders. The plants flower when the gull colony is in full breeding clamour, each plant producing one to three dinner-plate sized inflorescences.

And a 'dinner-plate' is exactly what the inflorescences are as far as the eager swarms of blowflies are concerned. Above the floral chamber the spathe flares out into a horizontal platform, mottled greenish-grey with flesh-coloured veins and streaks. It is covered on the upper surface with dark red hairs, and the overall effect, to the human eye at least, is an astonishing resemblance to a slab of rotting flesh. In addition, the spadix and the surface of the spathe release a stench which, in both quality and quantity, one would expect from the well-matured corpse of something the size of a sheep – quite remarkable for a mere 38 centimetres (15 in) of arum lily inflorescence. The first morning the plant unfolds its magnificent spathe, the blowflies, mainly gravid females, must choose between the arum lily and the real carrion that is present in the gull colony.

On the basis of odour alone it is a one-sided contest. Attracted by the irresistible aroma, the flies zero in on the spathe. They land, exploring its carrion-like surface for food and a place to lay their eggs. Although genuine putrefying flesh is absent, the flies are not deterred. Real carcasses frequently have their skins intact and only persistent searching by the flies will find an orifice – an eye-socket, mouth or anus – leading to the putrefying interior. So the flies on the spathe of *Helicodiceros* persevere, searching for a dark, damp and particularly foetid aperture. Everything about the spathe leads them to the neck of the chamber; the dark patterns, the density of the hairs, the direction in which they point, and the concentration of the smell. Behaving as they would on a real carcass, the flies force their way into the neck past the spadix and down into the pitch-dark chamber itself.

The chamber is tall. The spadix rises from the centre of the floor and disappears through the roof. It bears a battery of horizontal mauve spikes, below which there is a whorl of tightly packed female flowers, while the male flowers are borne in a ring above the spikes. Above the male flowers, the spadix, armed with 30 or so upward-pointing tapered fingers, curves out through the neck of the chamber, on to the flat surface of the spathe.

The female blowflies cannot escape from this damp and putrid-smelling space – the smooth fingers and hairs jam the neck of the chamber. The high humidity and the rotting-flesh smell excite some of them to the point of ovipositing, and piles of eggs mound up on the floor of the chamber. Many of them hatch into maggots, but these are doomed to perish for lack of food. Not so their parents, for the female flowers produce nectar, which sustains the prisoners and lures them on to the receptive stigmas so that pollen transfer can take place. Their detention for the arum lily's pleasure is not a one-day pleasure trip, however. Only when the female flowers have ceased to be receptive, about a day after the flies first enter the chamber, do the male flowers start to produce pollen. The flies must therefore be detained for long enough to receive this new pollen, which means two to three days of captivity. Not all of them survive this long. Sometimes the inflorescence kidnaps flies in such quantities that they clog the chamber and suffocate. But on the third day of imprisonment the hairs that block the neck of the floral chamber wither, and those captives still fortunate enough to be alive clamber out into the Sardinian sunlight, groggy and coated with yellow

97

pollen. Some of them, short of memory perhaps, or just desperate to lay eggs, will once again fall victim to the tempting aroma of a newly opened *Helicodiceros muscivorus*, so bringing about cross-pollination.

Other 'murderers'

In Japan a little arum lily, *Pinellia*, relies for its pollination on the numerous swarms of small gnats (Ceratopogonidae). The floral chamber is divided into upper (male) and lower (female) compartments, with a constriction between the two. To begin with the gnats tumble straight past the constriction into the lower chamber. At this stage, the female flowers are receptive, and will be fertilized by any gnats carrying *Pinellia* pollen. As there is no escape from this chamber, these early visitors perish. Later, when the male flowers in the upper chamber mature, pollen rains down and partially clogs up the constriction. Late-arriving gnats automatically pick up this pollen as they tumble past to join their dead comrades below. But they do not share the same fate, at least not in *this* inflorescence, for an exit hole now forms in the wall of the lower chamber by the action of a special swelling tissue. The timing is perfect – for the arum lily that is. For the lucky gnats that escape from one *Pinellia* inflorescence, there is always the chance that a visit to a second one might prove fatal.

Some arum lilies of the genus *Arisaema* are also killers. *Arisaema triphyllum* (jack-in-the-pulpit) is said to mimic the odour of fresh fungi and its visitors are male and female gnats intent on mating and egg-laying. In contrast to most other arum lilies, jack-in-the-pulpit is dioecious: the male and female inflorescences are borne on separate plants. In both sexes, the floral trap remains active for long periods – in the males for several weeks. Every evening, gnats assemble around the spathe, land, and tumble into the chamber below. They blunder about in an effort to escape and quickly become dusted with pollen. As the male inflorescence is equipped with an escape hole, it is not long before they are free again. Not so lucky are those gnats that fall into a female inflorescence, for although very similar to the male one in construction, it lacks an exit. The visitors must die. Some of them may have been carrying *Arisaema triphyllum* pollen. If so, their demise will not have been in vain.

The arum lilies are not the only plants with murderous flowers, however. Perhaps the most surprising – and beautiful – of all flowers that destroy their pollinators are those of certain water lilies such as *Nymphaea capensis*. Large, richly coloured and sweet-scented, the flowers draw humans and insects alike with their handsome advertising. Insect visitors include hoverflies, bees and beetles eager to gather the pollen that is offered in profusion by the flowers in the male stage of their development. The stamens, like sugar-coated lollipops, crowd the centre of the radially symmetrical flower, which opens for three or four days in succession. Each night, or during the day if the temperature drops too far below tropical levels, the flower closes. Each morning, when the flower reopens, a new batch of stamens has ripened, offering pollen galore. If any lesson is learned by the visitor it is that water lilies are an abundant and reliable source of food.

But this picture of endless bounty is no more than a façade. For on the first day of its life, the flower, then in the female stage, carries no pollen on

Above left A South African water lily, *Nymphaea capensis*. On the first day of its life the flower is in its female stage and may trap and drown several flies, bees and beetles (*above right* and *centre* – flower in section), though it requires only one with pollen on its body to effect pollination. That evening the flower closes; a beautiful, fragrant, floral sarcophagus. When it opens the next morning (*below right*), the stigma is covered over by the stamens and pollen-seeking insects may now visit the flower in complete safety.

99

its stamens. Instead, they stand like a glistening palisade around a circular pool of liquid in the centre of the flower. The bottom of the pool is the stigma itself, flat, circular, and totally different in design from the elongated female organs that protrude from the centre of most flowers. Visitors from the older lily flowers seem not to notice the slightly different architecture of the first-day flower. They land on the bloom looking for pollen, but there is none. They search, teetering around on the top of the stamens that surround the central pool. Instead of a sticky mass of pollen grains, the visitors are presented with a smooth waxy surface that is devoid of grip. If an insect ventures even half-way over the inner edge of the palisade, its feet slip and it tumbles into the liquid, which contains a wetting agent. For a second or two the victim struggles vainly to clamber out of the pool, but the stamens offer no foothold. The wetting agent allows even the lightest bee or fly to sink beneath the surface, and it quickly drowns. Pollen adhering to its body is washed off, sinks on to the stigmatic plate and germinates. The next morning the lily wears a kinder face.

Flowers that deceive by mimicking the prey of predatory insects

The larvae of most species of wasp, and those of certain types of fly, feed upon the bodies of living animals. The warble fly, for instance, lays its eggs on the hides of cattle. When the grubs hatch they penetrate the skin and form grape-sized cysts just under the hide, draining the cattle of health and energy, and causing much economic damage. Many species of wasp parasitize other creatures, laying their eggs in the bodies of caterpillars or stocking their brood cells with the paralysed bodies of spiders, caterpillars or crickets. Here is another opportunity for natural selection to secure an exploitative pollination system, provided it can equip a plant with a flower that mimics the living prey of one of these parasitic insects.

While it may be beyond the realms of nature's ingenuity for a flower to mimic an Aberdeen Angus, scientists have discovered an orchid, *Epipactis consimilis*, whose flowers apparently bear structures that resemble greenfly. The flowers are visited mainly by hoverflies, and occasionally by honeybees and solitary bees. Although the study area contained many species of hoverflies, only seven species were found visiting the flowers of *E. consimilis*, all of them having larvae that eat aphids. The behaviour of these hoverflies is most interesting. Males were seen hovering around the clumps of orchids, attacking other males and attempting to mate with any females that approached. The males also visited the flowers to feed on nectar, which is present on the lower part of the labellum. The flower has a jack-in-the-box device – triggered by the labellum – that catapults the fly on to the sticky stigmatic surface. In its efforts to break free, the fly brushes against the viscid tip of a pollinium, which becomes attached to its back. Sometimes male hoverflies were seen carrying whole bundles of pollinia, the result of visits to several different flowers.

The females behave rather differently, though the end result is the same – they are hurled on to the stigmatic plate like the males. The females hover tentatively in front of the flower, as if inspecting it, then land on the labellum and lay an egg on the flower. It seems as if the female hoverflies, fooled by the patterns and bumps, and possibly by the smell and texture of the labellum, 'believe' they are on an aphid-infested plant. If not, why are eggs laid in the flower, even when there are no aphids on the plant at all? *Epipactis consimilis* appears to be backing two pollination systems simultaneously: female hoverflies visit the flowers to lay eggs on what they mistake for an aphid-infested plant, and both the males and females visit to feed upon the sugary excretions of the labellum. It is interesting to note that were the latter the primary attractant, one would expect to find many different hoverfly species in the flowers, in addition to those whose larvae devour aphids.

Look-alike mates

No less fanatical than the female's urge to provide for her young is the male's urge to find a female. Courtship rituals and territorial confrontations are among the most spectacular and energetic of all animal activities, and here again there seems to be an opportunity too good to be missed. So far, with the exception of *Coryanthes*, we have not described any pollination system that relies solely on male insects. It is time to restore the balance.

By now it will come as no surprise to learn that a few species of orchid have evolved pollination systems that make use of the obsession that males have with the opposite sex. The only avenue open is to evolve a flower that attracts the attentions of the male by dangling in front of his nose a close facsimile of a potential mate. As male insects are extremely choosy about

The beautiful orchid *Ophrys speculum* attracts a male Scoliid wasp (*Campsoscolia ciliata*) by a combination of a specific scent and the colour, shape and size of the flowers. Its furry labellum so strongly resembles a female of his own species that he tries to mate with it.

their sexual partners (they rarely, if ever, attempt to copulate with a female of a different species) the flower could not get away with a sloppy imitation. Male insects recognize their partners by combinations of highly specific scents, shapes, colours, textures and behaviour, and a flower must mimic several if not all of these vital elements if it is to be successful in its deception. Many insects have defined, sometimes very brief, breeding seasons, and the flower's blooming period must coincide exactly with the male's sexually active period – no mean feat as both insect and plant seasons are subject to unpredictable variations. Notwithstanding these obstacles, a number of orchid species have successfully coupled their reproductive strategy to the mating behaviour of a single species of bee or wasp.

Take wasps of the family Thynnidae, for instance; a fascinating and unusual group that specializes in parasitizing the larvae of Scarabaeid beetles. The species of scarab parasitized by the Thynnids are mostly root parasites, and to find them the female wasps have to dig. This fact has dramatic implications for both the wasps and the orchids. For the wasps, it has meant that the females have abandoned their wings, exchanging the ability to fly for an improved digging performance – wings being a hindrance when tunnelling around in loose soil. But increased efficiency underground has been bought at the expense of top-side mobility, for the females can no longer fly in search of food and mates. The feeding problem is accommodated by the females' ability to suck the body fluids of their victims. And when it comes to mating, it is common in most Hymenoptera for the males to take the active role anyway. The females usually sit tight in an accessible position, release a pheromone, and wait. The males fly up the scent gradient until close enough to recognize the females by sight. The final coupling requires a combination of chemical, visual and tactile stimuli. For female Thynnids, therefore, the lack of wings is no great hindrance, but they have to climb to a good vantage point, usually a foot or two off the ground, before releasing their mating scent. The elegant winged males, constantly on aerial patrol, are quick to respond. But then a most unusual thing occurs. The male flies up to the female, who is clinging to a grass-stem or small branch in a head-up position, her jaws gnashing in anticipation. He then literally wrenches her off her feet. Like a bomber with an oversized missile slung beneath its fuselage the male flies off with his mate firmly in his clutches, her jaws clamped round his throat. They actually copulate in flight – a seemingly bizarre and needlessly hair-raising way of going about things. After all, though many bees, flies, butterflies and dragonflies mate in the air, the Thynnids could have stayed on terra firma for the whole event. So, one must look for some advantage in the Thynnids' unconventional approach.

The wasps remain coupled for a long time, sometimes several hours, and they do not just fly about aimlessly; the male carries his partner from flower to flower, where they both feed, still coupled. This seems to be the significance of the mating flight. It enables the female to have a nectar meal – the only one of her life – perhaps affording her a much-needed energy boost for her life of strenuous digging, or possibly a way of speeding up the maturation of her eggs. The male then has the responsibility of returning the female to a beetle-rich area, though just how he chooses is not understood.

Somehow a number of different orchids, all found in Australia, have latched on to the unique mating behaviour of the Thynnids and, using

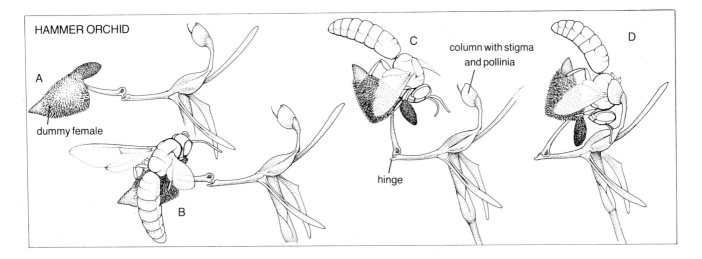

HAMMER ORCHID

A

dummy female

B

C

column with stigma
and pollinia

hinge

D

The labellum of the hammer orchid, *Drakaea fitzgeraldii* (A), bears a remarkable resemblance in terms of shape, colour and scent to the wingless female of one particular Thynnid wasp species. The male wasp of this species, fooled by the likeness, lands on the orchid's labellum (B) with an urge to mate. If he had found a true female wasp, he would have flown off with her grasped between his legs. When he attempts this with the flower (C), however, he is, instead, thrown back by a pivot mechanism on to the orchid's pollinia and stigmatic plate (D).

similar techniques, have turned it to their own advantage. They include the hammer orchids (*Drakaea*), the elbow orchids (*Spiculaea*) and the genus *Caladenia*, which contains the stunningly beautiful spider orchids.

In the hammer orchids, the labellum is modified to resemble the small, plump, wingless female Thynnid wasp. Even to our eyes, the resemblance is near-perfect, with a shiny head, a rounded, slightly hairy body, and an uptilted end to the abdomen. The deception even includes the scent – a copy of the pheromone released by the female wasp. The dummy female sits on the end of a short hinged arm and is free to bob up and down in the wind: hence the name, hammer orchid. In perfect line with the arm and its hinge is the column, where the orchid's stigma and anthers are housed. When a male wasp grabs his 'female' and attempts to fly off with her, his momentum carries him and the dummy up and over, pivoting on the end of the hinged arm and bringing his back smartly into contact with the orchid's pollinia and stigmatic plate. The whole action may be over in less than a second, but in that time he has already started to probe the lower end of the dummy with his genitalia. It is only when he finds his 'mate' firmly fixed to her plant that he gives up and departs. Each of the four species of hammer orchid is pollinated by a different species of Thynnid wasp. Pollinator specificity is maintained by the scent given off by the dummy female, which is only attractive to males of the correct species.

How could such an apparently foolproof system fail? Very simply. In competition with the real thing, the dummies are a miserable failure. If the orchids are in flower when the genuine female Thynnids emerge from the soil, the males hardly visit the flowers at all. But even this is not a fatal flaw in the hammer orchids' strategy. They make use of a final refinement of the Thynnid wasp life-history – the fact that the males appear in spring several weeks *before* the females. It is this frustrating interlude in the life of the male Thynnids that the hammer orchid puts to good use. By producing flowers before the female wasps appear, the orchids have the attentions of the males to themselves.

The term pseudo-copulation has been coined to describe this pollination strategy and it took a twentieth-century biologist, Monsieur A. Pouyanne, 20 years of observation and experimentation before he dared venture into print with the suggestion that male insects copulate with flowers. For

many years Pouyanne was President of the Tribunal of Sidi-bel-Abbès in Algeria, and he became interested in the pollination of various species of the genus *Ophrys*, which he observed growing on the ramparts of Algerian fortresses and on exposed railway embankments. His discoveries stunned and shocked him. Each species of *Ophrys* was visited by only one species of insect. And the visitors were all males. The flowers contained no nectar. From the actions of the males on the flowers Pouyanne concluded that they were trying to copulate with the labellum. In doing so, they transferred pollinia from flower to flower.

Keen to be totally sure of his ground before risking ridicule with so gross a suggestion, Pouyanne conducted numerous experiments. Using *Ophrys speculum*, he removed the labellum from the flowers, whereupon the usual visitors, male *Campsoscolia ciliata* wasps, neglected them. If he held the flowers upside-down so that the beautifully patterned upper surface of the labellum was out of sight, the male wasps approached but did not land on the flowers in the normal way. Female Scoliids seemed to find the flowers repugnant, but if Pouyanne concealed a number of flowers under sheets of newspaper, male wasps approached and appeared to search the vicinity for the orchid flowers.

Each species of *Ophrys* orchid mimics the female of a different species of wasp or bee. Here a horned bee (*Eucera nigrilabris*) is in the act of pseudo-copulation with *Ophrys tenthredinifera*.

The flowers themselves are somewhat wasp-like to the human eye. The centre of the labellum is a beautiful iridescent bluish-mauve, resembling the sheen reflected by the wings of a wasp sitting at rest. The margins of the labellum are burgundy-brown and densely fringed with velvet hairs. Pouyanne observed that the male Scoliid wasp always squatted on the labellum with his head in contact with the column, his abdomen and extruded genitalia probing the hairs at the apex of the labellum. The conclusion was unavoidable – the labellum of *Ophrys speculum* was mimicking the female of *Campsoscolia ciliata* so expertly as to stimulate the male wasp to mate with it, thereby obtaining the wasp's services as pollinator.

Pouyanne, and more recently Professor Kullenberg, extended the study of the *Ophrys* genus to include at least fourteen species, each one pollinated, by pseudo-copulation, by a different species of insect. The deception is not always perfect however, and occasionally pollen from one species of *Ophrys* is transferred to the stigma of another – a case of mistaken identity on the part of the male insects – and hybrids are produced. In every case the labellum is the centre of attraction, mimicking the appropriate female insect by a combination of colour, pattern, shape, size, texture (velvety hairs or smooth reflective areas) and, of course, the all-important scent. The pollinators are Scoliid or dagger wasps, digger wasps (Sphecidae), mining bees (*Andrena*) and horned bees (*Eucera*). In contrast to the situation in hammer orchids, Kullenberg found that some *Ophrys* flowers emit a scent that is more effective than that of a genuine female in attracting the attention of its male pollinators.

While the *Ophrys* genus was puzzling European botanists, Edith Coleman was investigating another example of pseudo-copulation in Australia. Between 1929 and the early sixties she and her colleagues published a series of papers describing the pollination of orchids belonging to the genus *Cryptostylis*. At least four species, *C. leptochila*, *C. subulata*, *C. erecta* and *C. ovata*, are pollinated by male ichneumon wasps of the species *Lissopimpla semipunctata*. The flowers (known as tongue orchids because of their elongate, strap-like labellum, with its pinkish, warty appearance) attract male wasps by means of a scent that can compete effectively with a receptive female wasp. The male backs on to the labellum and probes the area of the stigma and the pollinia with his genitalia; sometimes seminal fluid is ejected on to the flower. The pollinia become glued to the tip of the wasp's abdomen, and when he attempts to copulate with another *Cryptostylis* flower his thrusting movements crush the pollen against the stigma. How strange that four species of *Cryptostylis* are pollinated by the same species of ichneumon wasp, while in the case of the *Ophrys* genus each species is served by its own distinct pollinator. To avoid pollen being wasted on the stigmas of the wrong orchid, with the additional risk of causing hybridization, one would expect each species of *Cryptostylis* to form a liaison with a different type of wasp, especially in the case of *C. loptochila* and *C. subulata*, whose ranges overlap. As no hybrids have been found, one must conclude that some sort of physiological barrier to hybridization must be in operation.

The most recent research on pseudo-copulation, carried out in Israel, brings yet another species of orchid, *Orchis galilaea*, into this category – or at least close to it. In recent studies, it was found that the only visitors to the flowers of *O. galilaea* were male Halictid bees (*Halictus marginatus*),

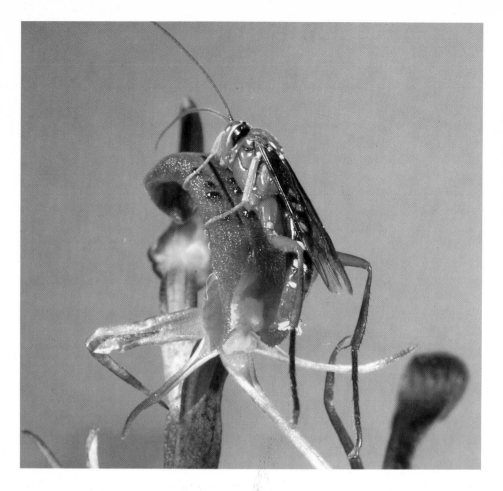

During his sexual encounters with a *Cryptostylis* orchid a male *Lissopimpla* ichneumon wasp has acquired a parcel of yellow pollen. Stuck securely to his abdomen, the pollinium will be transferred to another flower, when he is again fooled by the remarkable mimicry of this orchid flower.

despite the fact that the orchids were growing in areas where insects of all sorts abound. Perhaps more surprising was the observation that the flowers of *Orchis galilaea* were extremely variable in colour, and in no way resembled female Halictid bees in colour or shape. The flowers did give off a strong musky scent which the male bees obviously found attractive. Their inter-action with the flowers, however, was not predictable. Some approached upwind, made a pass around the inflorescence and flew off without landing. Others landed, but only briefly and without coming into contact with the orchid's reproductive parts. A few males, however, landed on the labellum, stood their ground for a few seconds, then moved into the centre of the flower where they wound up with the pollinia glued on to their heads. It seems safe to conclude that the flowers attract the bees via a sex-specific cue and that this is olfactory rather than visual. The term pseudo-copulation may not be quite appropriate, in view of the bees' tentative rather than whole-hearted behaviour. Maybe we have in *Orchis galilaea* a sort of half-way house be-tween normal floral attractions and the 'no holds barred' sex-dummy exemplified by the hammer orchid.

Pseudo-antagonism

Many animals defend territories and this is frequently done by males who are seeking the attention of females. The territory may be purely for mating, as in the spectacular lek behaviour of the American prairie chicken, or may provide the whole of the reproductive and food-gathering needs of a pair, as in European foxes. The holding of a territory for mating purposes only is also found in various Hymenoptera – bees from the genera *Eulaema*, *Centris* and *Anthidium*, for instance. In many cases the male bees are extremely pugnacious. Male wool-carder bees (*Anthidium manicatum*), common in many English gardens, frequently defend patches of flower-bed centred on one or two plants of lamb's tongue (*Stachys lanata*). The plant provides the bee with nectar and his females with plant-down, which they collect from the surface of the velvet-textured leaves and use to line their brood cells. A male wool-carder bee will defend his territory enthusiastically: bumblebees several times his size are rammed in mid-air and sent packing. Honey-bees about his equal in bodyweight, but with the advantage of a sting, are grabbed from behind, crunched with one swift squeeze from his powerful spiny legs, and dropped on the ground, mortally wounded or dead. If another male wool-carder bee appears on the scene, then a spectacular aerial dog-fight ensues.

In the Central American tropics, male *Centris* bees behave in much the same way. The bees' territories often include orchids of the genus *Oncidium* (*O. hyphaematicum*, *O. planilabre* and *O. stipitatum*), whose flowers are borne in long, slender racemes which dance and wave in the slightest breeze. Although the flowers do not always resemble insects, to the human eye at any rate, the male bees seem to treat them as insect intruders, flying at the flowers and butting them forcibly in the centre. The orchids' pollinia attach themselves precisely to the bee's forehead. When he strikes another flower they are crushed against the stigmatic plate.

There is more to this relationship than meets the eye, however. *Centris* bees are fiercely loyal to their territories: even captured bees return to them when released. They are not in the habit of wandering off through the forest attacking *Oncidium* flowers at random. It seems, therefore, that the best a male *Centris* bee can do for an *Oncidium* growing in its territory is to fertilize it with its own pollen, which rather makes a nonsense of the flower using a pollinator at all. There is, however, a final twist to the conundrum. Female *Centris* bees pollinate the flowers of *Oncidium ochmatochilum* and the related *Odontoglossum grande*, though their behaviour is not the flying attack of the males. Instead, the females land and then attempt to push their way into the centre of the flower. It is not known whether these orchids provide nectar or not, or whether the same flowers are subjected to attack by males. However, many *Oncidium* flowers bear a close resemblance to flowers of the family Malpighiaceae, which just happen to be the favourite source of nectar or oil for female *Centris* bees! Perhaps certain *Oncidium* species are running two parallel pollination systems – their flowers are self-fertilized if they happen to be in a male *Centris* territory, and if not, they rely upon their similarity to the Malpighiaceae to attract a female *Centris* bee.

We will have to await the outcome of further research. But I venture to submit that the 'unacceptable' face of pollination is not so unacceptable after all if it leaves us with the picture of a dappled jungle clearing, an orchid with fragile dancing flowers, and a frenzied little furry bee with pollinia stuck on his forehead.

CHAPTER SIX

Pollination without animals

So far in our story, animals have definitely been the heroes and heroines of pollination because they tend to promote outbreeding, which may lead to the creation of new, genetically superior plants. The pollen they collect on one flower often goes to another flower belonging to the same species but found on a different plant individual. The various devices plants possess to prevent, or at least minimize, self-pollination so impressed Charles Darwin that he came up with his famous dictum that 'nature seems to abhor perpetual self-fertilization'.

However, cross-pollination is, of course, downright impossible for a plant individual that is doing a pioneering job – one that has invaded a new environment all by itself, far away from the plant partners it could normally have relied on. Weeds often act as pioneers, and the fact that many of them are self-pollinators, or 'selfers', is not surprising at all. But the price they have to pay is genetic uniformity. Even for plant species that grow in groups, it seems a wise strategy to have the option of self-pollination available when insects are scarce, inactive or, as in a hostile environment, totally absent. How wise is best illustrated by the European bee-orchid (*Ophrys apifera*). From simply looking at the flowers, which are beautiful mimics of bumble-bees, one might imagine that the plant reproduces by means of pseudo-copulation, the method which (as we have seen) works so well in some other members of the *Ophrys* genus. Wrong! The insect that supposedly 'corresponds' with the bee-orchid has not been found, and the plant habitually reproduces by 'selfing'. It may be that the pollinator existed in the past, but was wiped out by adverse environmental conditions such as occurred in the Ice Age, and that rather than follow suit, the orchid developed a self-pollination mechanism that worked. Another neat example is found in Hawaii, where certain tree-like lobelias have become habitual selfers after the birds that pollinated them died out in the recent past. Self-pollination is also the pollination method used by the 138 species of low-growing flowering plants that live in Timbuktu. In the daytime, the ground here is so hot that insects simply don't stand a chance; they do their work higher up in the canopy, where it is cooler.

A number of plants that normally depend on cross-pollination, such as the well-known nasturtiums, also have back-up self-pollination systems: the filaments of the stamens carry out certain movements during the period the flower is open, the end result being that the anthers will touch the still-receptive stigmas if pollination has not already occurred. In other instances of back-up self-pollination systems, the anthers in the young flowers are placed below the stigma, which they reach by simple elongation of the filaments later. Movement of the style, as demonstrated by fireweed, can also

Species using the wind as a pollination agent tend to produce large quantities of dry, dust-like pollen which readily takes to the air as their flowers sway in the breeze. It has been estimated that the average single floret of rye grass (*Lolium*) produces about fifty thousand grains.

109

bring about such contact. In some species of coffee, such as *Coffea liberica*, the anthers shed their pollen while they are still enclosed in the bud. When it opens, the stigmas, which at that time have just become receptive, are already covered with the precious dust. In foxglove, the whole corolla, including the attached stamens, is shed at the end of the flowering period, but in the process the anthers may very well touch the stigma and achieve pollination, if it has not taken place earlier. The most extreme case of self-pollination, however, is provided by flowers that never open – the so-called cleistogamous flowers.

Cleistogamy

Cleistogamy (from two Greek words together meaning a secret marriage) was discovered as early as 1732 by Johann Jakob Dillenius, who seems to have used it as an argument against the sexuality of plants. Carolus Linnaeus, that pioneer classifier of things botanical, was already familiar with at least a dozen different species that showed the phenomenon, and a great many more have been added to the list since. Plant species with cleistogamous flowers are distributed over a wide and quite diversified range of plant orders, and well-known examples are violets (*Viola* spp), a rush (*Juncus bufonius*), wood sorrel (*Oxalis acetosella*) and a species of rice (*Oryza clandestina*). Practically all cleistogamous plants can also produce normal or chasmogamous flowers (from the Greek word *chasma* for opening): the grass *Festuca danthonii* is a rare exception.

Environmental circumstances have a great deal to do with the formation of cleistogamous flowers. Linnaeus already knew this: in the cold Swedish climate where he lived, some species of rock rose (*Cistus*) and sage (*Salvia*) which he had obtained from the warm Mediterranean region produced only closed, self-pollinating flowers – in contrast to their normal habit in their native Spain.

It was found that in several instances day-length was the decisive factor. Plants of *Viola cunninghamii*, for instance, when exposed to long days, will form only cleistogamous flowers; kept on a short-day regime, they produce only chasmogamous ones. Other violets, such as the sweet violet (*Viola odorata*) and the dog violet (*V. canina*), probably respond in the same way to day-length; at least, that is suggested by their behaviour during the growing season. In early spring they produce large, normal flowers which are pollinated by insects with fairly long tongues, such as the beefly (*Bombylius*), honey-bees and bumblebees, various wild bees, and butterflies such as the fritillaries, whose caterpillars live on the violets' leaves. In late spring and summer, small cleistogamous flowers put in their appearance. They owe their smallness to the fact that all the non-essential organs in them have been suppressed: the corolla lobes appear as minute scales, and the calyx lobes are only half-size when compared with those of normal flowers. There are only two stamens in the cleistogamous flowers instead of five, with only two pollen sacs each instead of four. However, the few pollen grains that are formed are quite normal, and so are the ovules in the ovary. Germination of the pollen grains takes place while they are within the anther wall which separates them from the adjacent stigma. The fruits formed in these cleistogamous flowers are normal in appearance and they tend to mature a little faster than those from the chasmogamous flowers. The seeds, likewise, are quite normal and are eminently viable.

The advantages of cleistogamy as a survival strategy of jewel weed (*Impatiens capensis*) have been beautifully documented. As an annual whose seeds do not display much dormancy, jewel weed is absolutely dependent on efficient pollination: it has to be achieved – no matter what. Moist, well-illuminated sites provide optimal growth factors and it is here, between March and September, that 2 metre (6 ft) tall jewel weed plants appear, each capable of producing many showy and fairly big, long-spurred chasmogamous flowers which offer a rich reward of nectar to visiting bumble- and honey-bees, hawkmoths and hummingbirds. Unfavourable sites are those with well-drained soil and little light. Here, the jewel weed plants remain stunted. They may flower very early in May, but will then produce only cleistogamous flowers, and they die soon afterwards. But despite all the adversity, populations may maintain themselves in such unfavourable sites for years. The main advantage of cleistogamy is that it is a cheap method. Producing and maintaining large, nectar-rich chasmogamous flowers is biologically expensive, the cost of producing a seed through cleistogamy being only two-thirds of what it is for one formed through the normal method. Small wonder then, that a plant will resort to cleistogamy under unfavourable circumstances. Another interesting point is that well-developed jewel weed plants that would normally form a number of chasmogamous flowers, will make only cleistogamous ones after they have been grazed by deer, or when the ends of their branches have been cut off – a quick response to an emergency situation.

Pollination by the wind

When told that grasses are flowering plants, many people are quite surprised. With few exceptions, grass flowers are inconspicuous in terms of both colour and smell. Moreover, they do not embark on the striking and fanciful relationships with insects or birds which many of us have begun to appreciate. Some resort to self-pollination, but by and large they are pollinated by the wind.

An important point to note at this stage of the story is that wind pollination, or anemophily, is by no means confined to grasses: it is the common mode of pollination in sedges, rushes, nettles, coniferous trees such as pine, fir and cedar, and broadleaved ones such as walnut, oak, birch, poplar and hazel. At first sight a motley crowd – a heterogeneous bunch of herbs, shrubs and tall trees. What do all these plants have in common? It is significant that they have a tendency to form large communities with individuals belonging to their own species, or to a few species only. Good examples are grasslands, and the vast coniferous forests of the Pacific Northwest in America. Such communities are marvellously suited for wind pollination; as the breeze passes over the flowering grasses in a meadow, or through the canopy of trees in a forest, it may release or dislodge whole clouds of pollen grains which are bound to effect cross-pollination somewhere as they are carried along by the wind.

Another common characteristic is what we might call an assembly of negative features. The typical wind-pollinated flower is small, and of a greenish or nondescript colour. It has neither nectar nor scent, is constructed in a very simple manner, and lacks petals; the flowers may even be totally naked, as we see in poplars and ashes. Not surprisingly, adaptation to wind pollination is usually accompanied by separation of the sexes, which is an elegant device to minimize self-pollination. Many conifers such as spruce, pine and silver fir, a majority of sedges (*Carex*), alder, hazel, walnut, oak, beech, and maize (Indian corn) are monoecious – separate male and female flowers are found on the same plant. Dioecism, with separate male and female plants, is found in hemp, hops, the large nettle (*Urtica dioica*), mercury (*Mercurialis*), poplar, yew, juniper, and a number of palms such as the date palm. However, a number of wind-pollinated plants – most grasses, rushes and pondweeds – have hermaphroditic flowers with both functional stamens and pistils. The chances are that these plants have other barriers against self-pollination: chemical barriers, or dichogamy, in which stamens and pistil do not reach maturity at the same time, are found in meadow rue (*Thalictrum*), pondweed (*Potamogeton*), plantain (*Plantago*), and some species of sorrel (*Rumex*).

An evolutionary 'change of mind'

The haphazardness and waste that characterize anemophily have led some botanists to believe that wind pollination is a primitive phenomenon. In plants such as the conifers, the maidenhair tree, and in cycads – in short, in the gymnosperms, which are generally conceded to be primitive – wind pollination is very common. Moreover, the fossil record indicates that the ancestors of the conifers have always been wind-pollinated as well. However, some other gymnosperms are insect-pollinated. According to one authority *Cycadoidea*, a possible ancestor of the flowering plants, was beetle-pollinated. Among the living cycads, *Encephalartos* – a very ancient genus with several species in South Africa – is also beetle-pollinated.

With almost palpable enthusiasm, Porsch in 1910 described the situation in *Ephedra campylopoda*, a plant of the Adriatic region. Some plants of *E. campylopoda* bear female flowers only, while others have male and female flowers mixed in one inflorescence. The paired female flowers have bracts of a very vivid yellow colour, which turns red later. They attract numerous insects, including honey-bees, which also visit the male flowers, and there can be no doubt that these insects do pollinate. A reward is found in the drop of sugary liquid which protrudes from the end of the tube formed by the outer cover of the ovule. The drop also acts as a stigma, catching the pollen grains. In *Welwitschia bainesii*, the famous two-leaved 'living fossil' found only in the Namibian desert of Southwest Africa, a hemipterous insect, *Odontipus sexpunctatus*, has been claimed to carry pollen from the male to the female plants. Porsch challenged this, but he was later able to provide evidence that small bees and flies do the job. Again the small drop of sugary fluid which oozes from the end of the covering of the ovule serves as both reward and stigma. It is findings such as these that lead many botanists to believe that, in the course of evolution, a very gradual change took place in certain gymnosperms so that they became insect-pollinated, and that perhaps all the early flowering plants that arose as their descendants were insect-pollinated also.

112

These curious masses of twisted leaves belong to *Welwitschia*. It is, in fact, a gymnosperm bearing up to twenty scarlet female cones on long stalks, each scale of which holds one broad-winged seed; its male cones are similar. Observations have shown that *Welwitschia* pollen is occasionally transported by insects – a characteristic more usually associated with higher plants.

The change in character of the pollen required for the successful transition from wind- to insect-pollination was quite profound. Wind-borne pollen must be dry and not sticky, otherwise it tends to form clumps, making dispersal very difficult. On the other hand, for an insect-borne pollen, stickiness is a most desirable characteristic, and we do indeed find that such pollen is coated with a glue formed by the anthers. In general, this glue is a thick, sticky, non-volatile oil which retains its fluidity for quite a while. The discovery, in 1920, that the pollen grains of a number of wind-pollinated plants in the buttercup family – generally accepted as primitive – display traces of the oil was, frankly, rather sensational: it provided strong support for the idea that, in these particular flowering plants at least, wind-pollination is a secondary feature, a return to the alliance with the wind which had been so successful in their distant ancestors.

The fossil record indicates that such a 'change of mind' of some insect-pollinated flowering plants may have begun as early as the Middle Eocene, some 50 million years ago, and must still be going on today. Some families contain genera and species that are insect-pollinated, while others are wind-

pollinated. In the willow family (Salicaceae), for example, poplars are pollinated by the wind, while many willows are pollinated by insects. In the dock
or buckwheat family (Polygonaceae), buckwheat and bistort are insect-
pollinated, and docks and sheep-sorrel are wind-pollinated. Even species
belonging to the same genus may have different modes of pollination.
Common mugwort (*Artemisia vulgaris*), a weed of roadsides and waste
ground, has a dry and powdery pollen that is wind-dispersed; *Artemisia
glacialis*, found at high altitudes in the western Alps, has sticky pollen
carried by insects. The change from insect-pollination to wind-pollination
has been beautifully documented also in the maples: sycamore (*Acer pseudo-
platanus*), Norway maple (*Acer platanoides*) and common maple (*Acer
campestre*) are all pollinated by insects, while the pollen of box elder (*Acer
negundo*), with its nectarless and corolla-less flowers, is dispersed by the
wind.

Wind pollination and pollination by animals may even occur regularly
within one plant species. This is true for ling (*Calluna vulgaris*) and certain
species of plantain (*Plantago*) which, very early in the morning, are visited
by pollen-eating hoverflies. The most striking case of such 'double-barrelled
pollination' is found in the sweet or European chestnut. During the first
part of the flowering period the pollen grains are sticky and the inflorescences
attract numerous insects (mostly small beetles). Towards the end of the
flowering period the pollen becomes dry and powdery and is dispersed by
the wind.

The numbers game in wind-pollination

Slight shaking of a flowering twig of hazel or pine brings forth a cloud of
pollen – a neat demonstration of how intrinsically wasteful the saturation-
bombing practised by wind-pollinated plants can be. In areas where coniferous trees abound, there may be so much pollen on the ground that some
people have seriously considered the possible existence of 'sulphur-rains'.
Conifer pollen regularly forms a conspicuous scum on the water of Crater
Lake in Oregon, while on glaciers in the Pacific Northwest, abundant
deposits of pollen provide food for very primitive wingless insects called
Collembola. It is thus perfectly legitimate to talk about an aerial pollen-
plankton. And yet, strange to say, an elaborate comparison of wind-
pollinated and animal-pollinated plants shows that the former are not con-
sistently better at producing pollen.

There is a very wide range in pollen numbers in both categories, resulting
in a tremendous overlap. Among the best wind-pollinated performers are
sorrel (*Rumex acetosa*), with about 400 million pollen grains per inflorescence,
and lesser reedmace (*Typha angustifolia*) with 175 million. However, horse
chestnut (*Aesculus hippocastanum*), which is insect-pollinated, sports a very
respectable 42 million per inflorescence. Even when we look at the production of pollen grains *per plant*, hard and fast conclusions are difficult. It has
been estimated that an individual hazel bush may produce 600 million
pollen grains. But insect-pollinated corn poppies (*Papaver rhoeas*) come
close with 300 million grains per plant. The picture is undoubtedly complicated by the fact that so many animal-pollinated plants also use pollen as a
reward for their visitors. It is well to remember here that a single baobab

Billions of surplus, buoyant pine pollen grains form a scum on this lake in the American Rockies. Although seemingly wasteful, this excess pollen is by no means an indication of an unsuccessful pollination mechanism.

tree flower (*Adansonia digitata*), which caters to bats, may contain up to 2,000 stamens. What really counts, therefore, is not the absolute number of pollen grains produced per stamen, per flower, per inflorescence, or even per plant, but the ratio between that number and the number of available ovules. It ranges widely, from about 6,000 in *Polygonum bistorta* (insect-pollinated) to 2,500,000 in hazel (wind-pollinated). By giving just a little thought to the logistics of the various cases, one can easily see that it is a disadvantage for a wind-pollinated plant to have numerous ovules in one ovary. Indeed, the number actually found is one, or a very few.

It is amusing, for a moment, to think of plants as human beings accustomed to reasoning things out. They 'know' that ovules represent an expensive investment, and that each plant individual can afford to make only a limited number of them. The question then is, how to present these ovules: all of them together, locked up in a single ovary, or as individual ovules, spread out over a number of pistils or ovaries? The surface area of the receptive stigma at the tip of the pistil is extremely small in comparison to the total surface area of the plant and its immediate environment. In spite of the tremendous numbers of pollen grains released, the chance of a stigma being hit by a solitary pollen grain is, therefore, very small indeed, even if we consider just one hit. If we assume that chance to be one in a thousand, the chance of a pistil being hit twice amounts to one in a million. It simply does not make sense to put more than one or two ovules into each ovary: any more would just be wasted. The strategy of having many pistils with only one ovule, but with a large receptive stigmatic surface, is clearly greatly superior.

115

In order to gain the greatest chance of being dusted with wind-blown pollen, each of the female flowers of a walnut (*Juglans regia*) carries a large, lobed receptive stigma. Its ovary consists of two united carpels with a single erect ovule, which, when fertilized, will develop into a one-seeded fleshy drupe.

It is fun to check this against well-known examples. Everyone knows that hazelnuts normally contain only one seed; to find two seeds per fruit is such a rare event that one likes to share them with a friend and make a wish. Indeed, whole families that are typically wind-pollinated, such as grasses and sedges, nettles, plants in the goosefoot family (Chenopodiaceae) and members of the dock family (Polygonaceae) have an ovary with only one compartment containing a single ovule. The female walnut flower also has only a single ovule. Birches and alders (Betulaceae) have two ovules, one in each of the two compartments of the ovary. Ash (*Fraxinus*) also has two compartments, but each contains two ovules. Beech and oak (Fagaceae) have three chambers with two ovules each, but the fruits that develop from the ovaries are one-seeded, since five of the six ovules abort. It is very instructive to compare these situations with, for example, those found in the orchid family, which are typically insect-pollinated and possess a single-chambered ovary that may contain thousands of ovules (a million in the case of the vanilla orchid). An animal pollinator delivers a large number of pollen grains to the stigma in one fell swoop and therefore it is advantageous for a pistil to have many ovules, all waiting for the pollen tubes to arrive. Even more instructive is a comparison between poppies (*Papaver*), clearly insect-pollinated, with thousands of tiny seeds per fruit, and *Bocconia frutescens*, which belongs to the same family but is wind-pollinated and has only one seed.

How and when the pollen grains are liberated

Most wind-pollinated plants release their pollen only when conditions are favourable, and even then not all at once. The broadleaved trees that are wind-pollinated, such as alder and hazel, produce their flowers very early in the spring, when the tree is still bare, so that the pollen cannot be intercepted by their foliage. Since rain has a disastrous effect on the viability of

Thread-like filaments suspend the anthers of grass flowers by the centre of their two compartments (X-shaped configuration), allowing maximum wind disturbance during which pollen grains will be shaken free.

most pollens, and would also wash it out of the air, pollen should be liberated only in good weather when the air is dry. On top of this, we find that plants display a remarkable regularity in their pollen-release, so that it becomes available at a particular time of the day. In alder, hazel, birch and pine it happens around 2 p.m., when the temperature usually is at its highest and the air at its driest; an additional advantage is that a light breeze often puts in an appearance at that time. Grasses produce pollen much earlier in the day, but the exact time varies from species to species – a device that limits the chance of hybridization. Barley is somewhat unusual in that it has two periods of pollen-release, most of it being liberated between 7 and 8 a.m. and the rest between noon and 4 p.m.

In many grasses, the emergence of the stamens from the flowers is almost explosive. The filaments stretch in a spectacular manner, reaching their full length in about ten minutes. The two compartments of each anther touch each other only in the middle and diverge from each other towards their ends, so that they form an X-shaped figure. The filament is attached to the midpoint of the X, and this gives exceptional mobility to the anthers, which sway back and forth in the slightest breeze. In doing so, they may shake out parts of their contents. The fact is that many grasses operate on the 'hour-glass' principle. In good weather, a hole is formed at the lower end of the anther, and in the absence of wind, the pollen that emerges through it simply accumulates on a small, protruding horizontal ledge at the base. When the wind finally comes, the pollen is blown away and the anther begins to swing, with the result that some more pollen will emerge from the 'hour-glass' – but only a little more.

A similar economy mechanism is found in hazel and some other catkin-bearers. Here, in good weather, a gaping hole forms over the entire length of the anther, whose volume is thereby reduced to less than a third of what it was. The pollen, having no place to go, must come out. In the absence of wind, it does not fall to the ground immediately, but simply remains spread

117

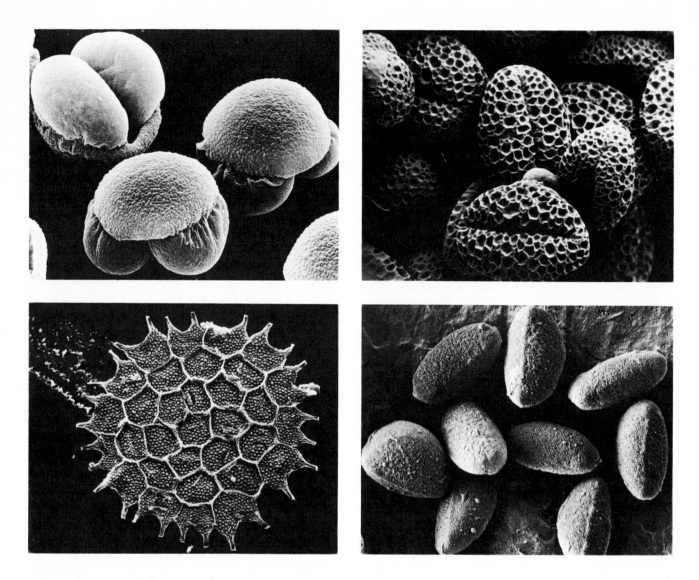

out on the parts of the catkin immediately below – until the breeze arrives. Stamens of hoary plantain (*Plantago media*) have a stiff filament with a very mobile anther at its tip which waves in the wind like a flag atop a mast.

Other wind-pollinated plants such as stinging nettles and some of their relatives, for example the Japanese paper mulberry (*Broussonetia papyrifera*) and the little *Parietaria* that is sometimes grown in greenhouses, have an explosive mechanism to shoot the pollen up into the air. Within the bud, the filaments of the stamens are bent inwards towards the centre of the flower, where they are held under terrific tension, like springs. When the bud opens the tension is released and the stamens suddenly spring outwards while the anthers split and discharge puffs of pollen. In *Parietaria*, the pistil reaches maturity earlier than the stamens and it is only after the star-shaped stigma has sloughed off that the mechanism comes into action. Again we see that self-pollination is prevented. Annual mercury (*Mercurialis annua*) goes one better than the nettles: it has separate female and male flowers, the latter with about 20 stamens each. At the right moment, the male flowers are literally catapulted away *in toto*. The anthers open simultaneously, releasing their pollen puffs.

There are no ironclad rules to the shapes of pollen grains, and they are, in fact, as unique to each species as fingerprints are to human individuals – electron-micrographs reveal an astounding diversity. *Clockwise from top left: Pinus sylvestris, Forsythia, Dombeya and Narcissus.*

What counts in a wind-dispersed pollen

Apart from the absence of glue on the pollen grains, it seems that a number of other requirements have to be fulfilled in order for wind pollination to be successful. It is, of course, crucial that the grains do not sink to the ground too fast, and this consideration requires that the grains must not only be minute but also very light. Strange to say, when we compare a large number of wind-borne and insect-borne pollens in terms of size and weight, no clear differentiation emerges. The best we can say is that the truly massive grains are found in insect-pollinated plants such as pumpkin (*Cucurbita pepo*), yet pumpkin pollen grains are about a thousand times as large as those of horse chestnut, which is insect-pollinated also, and these in turn are many times larger than the truly diminutive pollen grains of insect-pollinated forget-me-nots (*Myosotis*).

Perhaps the only real lesson which can be learned from all these measurements, then, is that stickiness, or the lack of it, is really the decisive factor. A dry, powdery, flour-like pollen in which all the grains stay separate is the ideal for successful wind pollination.

To enhance their aerial mobility pine and spruce pollen grains have air-sacs, giving them a Mickey Mouse appearance, and their presence counteracts successfully the fairly large weight of these grains. Larch has pollen grains that are at least twice as big as those of hazel, yet the air-sacs are curiously lacking. However, in this plant we find that the grains lose water very rapidly once they are exposed to air, and consequently their shape changes and they become shallow, round bowls. The water loss is not serious, and is remedied automatically when the grains hit an appropriate moist stigma.

The strange peregrinations of wind-blown pollen

Once it is in the air, pollen can be carried to a considerable height by ascending air currents. Above the Mississippi River Basin in America it has been shown to be present at an altitude of 5,800 metres (about 19,000 ft). Even if it goes up only to a modest 2,000 metres (6,500 ft), and the prevailing wind does not exceed 16 kilometres per hour (10 mph), the pollen may in a 24-hour period cover a horizontal distance of up to 400 kilometres (250 miles). Pollen of *Nothofagus*, a Southern Hemisphere genus of beech tree, and *Ephedra* has been reported over Tristan da Cunha, 5,000 kilometres (3,100 miles) from the nearest growth-site of these plants; and pollens of alder, birch, oak, beech, grasses, plantain and sorrel have been found over the mid-Atlantic.

The question of whether the pollen grains can still germinate after spending days up in the air, exposed to such unfavourable influences as high temperatures and ultraviolet radiation, is an interesting one and the answer is different for different plants. Pine pollen is pretty tough, and can survive exposure to a temperature of 41°C (105°F) for 24 hours. Unfortunately, most pollens are quite sensitive to ultraviolet radiation. During the night and also in rainy weather, when there is little or no ultraviolet around, the pollen tends to come down. The practical importance of long-distance dispersal is thus quite limited, and perhaps one should think in terms of metres rather than kilometres.

119

In the flowering plants, why wind pollination?

Having arrived at the end of our wind pollination story, we can no longer avoid discussing an intriguing paradox that lies hidden in it. Earlier in this book we suggested that the flowering plants owe their spectacular success on this planet in large part to the development of pollination mechanisms that involve animals. But if success is a yardstick for 'superiority', how can we explain the fact that in the course of evolution so many angiosperms have gone back to the original 'inferior' and wasteful method of wind pollination? Are plants exempt from the good old tennis rule that one should never change a winning game? The answer is: of course not. Some of them have just followed the rest of the prescription, which says that one should always change a losing game. We mentioned earlier in this chapter that reliance on insects or other animals becomes a losing game when these creatures are not around, or when a hostile environment keeps them from always operating in a way that would be beneficial to plants. It follows that careful comparison of a number of environments must go hand in hand with an evaluation of the pollination systems found in them.

In central Europe, a region of hills and mountains which still has its fair share of forests, it was found that only one-fifth of the total number of species of flowering plants is wind-pollinated. On the other hand, on a group of North Frisian islands just a few metres above sea level and subject constantly to the full force of the ocean winds, the proportion of wind-pollinated plants was close to one-half. Wind pollination is also high in arctic regions, being found in over a third of the plant species in Iceland and Greenland, where grasses and sedges are prevalent. Other plants, such as certain species of willow (*Salix*), are wind-pollinated in the Arctic, while their close relatives in temperate regions are unquestionably insect-pollinated, at least to some extent. Strange to say, the parallelism which in many ways exists between high latitudes and high altitudes is not found in the realm of pollination. Above 2,000 metres (6,500 ft) in the Caucasus, only one-ninth of the plant species is wind-pollinated; in the Alps, one-sixth. There is undoubtedly a connection here with the rich insect life in these mountains. In the majestic virgin rain forest of certain tropical regions, such as the Amazon basin, wind-pollinated plants are exceptional: the almost complete absence of air currents in the lower layers, combined with the absence of vast, continuous populations of a single tree species, does not favour wind pollination. On the other hand, insects, birds, bats and monkeys abound, and many of these have become efficient pollinators. The question which of the two is 'superior', insect or wind pollination? is thus – essentially – a futile one. In the final analysis, it is the environment that decides.

Water pollination

Although at first sight it may seem that it is a very far cry from wind-pollination to water-pollination, there is every reason to discuss the two in one and the same breath – with wind pollination first. In those cases where we are dealing with plants whose pollination takes place entirely under water (the hyphydrophiles), the selfsame principles apply that we found operating among the anemophiles or wind-pollinated plants: saturation-bombing, lack of direction, and an almost total dependence on chance. Only the medium is different – water instead of air. Furthermore, the borderline between water- and wind-pollination is very vague indeed. As we shall see, several species of aquatic plants (the ephydrophiles) demonstrate their terrestrial origin by sending their male and female flowers to the water's surface, where the wind is actually doing the job. They represent a complete reversal, for as we have seen an aquatic environment must have been the cradle of life on this planet, and it was only after a long and painful struggle that some organisms became fully adapted to life on land. To go back to an existence in water must have been a veritable *tour de force*. It is a matter of simple observation that the degree of success achieved in this enterprise has not been the same for the various plants and animals involved. It would be impossible for seals to spend their whole life in the water, and for sea otters it would be a dubious proposition at best. Whales, dolphins and porpoises are perhaps the only mammals fully adapted to an aquatic life.

Turning to those flowering plants that have gone back to the water, one can ask which ones are the sea otters and which the dolphins? The range found in the degree of adaptation is even wider than it is in the mammals. Some plants have gone only half-way and spend their entire life on the surface of the water, for example the duckweeds (*Lemna*). They cover entire ponds with a thin green carpet of tiny leaves and rely mostly on vegetative reproduction. This, undoubtedly, is an adaptation. In some species of duckweed, flowers are exceedingly rare. They are minute, yet their spiny pollen grains indicate that they are pollinated by insects. A *Lemna* species in India demonstrates its relationship with arum lilies by having small beetles as pollinators. It may be significant that many true arum lilies display tendencies towards an aquatic life also. They may even go further than the duckweeds – some species of *Cryptocoryne*, prized aquarium plants, have long, almost tubular 'trap-flowers' (actually inflorescences) that may be formed entirely under water. They are filled with air, and when they reach the surface the upper part unfolds; the small lid on the slender pitcher is then displayed as a 'flag', which is strongly scented and usually provided with a fringe of hairs that aids in the evaporation of the smell. The inside of the tubular pitcher is covered with tiny wax bodies and is extremely slippery. The pollinators, small fruit flies attracted by the odour, find no foothold and fall down to the bottom of the floral chamber, where they may achieve pollination – some 25 centimetres (10 in) below the water's surface but still in air. Many other submerged plants, such as the well-known carnivorous bladderworts (*Utricularia*), send their inflorescences up, completely out of the water. Pollination thus presents no special problems.

The situation is totally different for the fascinating Podostemonaceae, sometimes referred to as waterfall plants. Most members of this family live in the tropics, attached to rocks in fast-flowing rivers, even in rapids. In appearance, they are strongly reminiscent of algae, and the construction of their vegetative parts, and the chemical nature of their cell walls, is such that they can survive strong turbulence without damage. But what can they do about their sexual reproduction? To reconcile it with frantic turbulence is really too much to ask. Fortunately the regions where they occur are characterized by extreme seasonal fluctuations in the water level, and the plants take advantage of the dry season, when they are fully exposed to

121

air, to flower and produce seeds. They do so 'in a hurry' and on a massive scale. Depending on the particular species, pollination is taken care of by selfing, or by the wind, or by insects. For the seeds to have a long period of dormancy would, of course, be fatal; they would simply be washed away by the fast-rising waters before the seedlings had the chance to emerge and establish themselves. So the seeds germinate almost immediately, and the young seedlings attach themselves to the rocks with a glue that is said to be better than the one used by barnacles.

A lovely example of a plant that sends its flowers to the water's surface for pollination and fertilization is ribbon weed (*Vallisneria spiralis*), well known to fanciers of freshwater aquaria. The species has separate male and female plants. The flower buds produced by the latter each contain only one little flower, enveloped in a thin membrane. While the slender stem on which it is placed slowly grows up to the surface in a distinctive spiral fashion, the female flower remains closed. Once there it gradually unfolds, exposing a number of tiny white petals and three pistils that are split in two. (For such a small flower, these pistils are quite large.) At the water's surface, the ribbon weed flower now sits right in the middle of a neat little dimple – the result of surface tension.

The floral buds of the male ribbon weed plants open while still underwater. On a short common stem we can discern a whole bunch of tiny round pollen flowers which for the time being remain tightly closed. By the time the female flowers have spread out on the water's surface, the male ones are released, one by one, and they shoot upwards. Many are probably gobbled up by fish before they reach the surface as the protein-rich globules are highly nutritious. Those that arrive safely open up and float on the surface like tiny boats, their three canoe-shaped petals acting as pontoons, keeping the two stamens in an erect position above the water. Even when a gentle breeze moves the miniature boats across the surface there is little danger

Below and right Ribbon weed (*Vallisneria spiralis*) has separate male and female plants. The male releases free-floating flowers – pollen boats. Those that are not eaten by fish may drift towards a surface-borne female flower. If they reach the perimeter of the dimple created by surface tension around that flower, they will slide down into it, then to be catapulted on impact on to a receptive stigma.

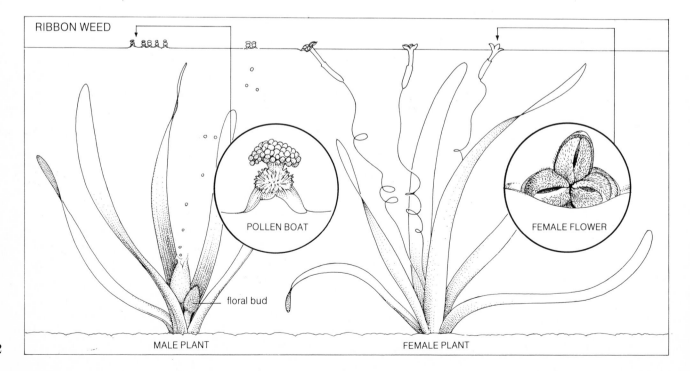

RIBBON WEED

POLLEN BOAT

FEMALE FLOWER

floral bud

MALE PLANT

FEMALE PLANT

that the clump of large sticky pollen grains in each anther will be wetted. Ultimately, a pollen boat's fate will be very much like that of Sinbad the Sailor's ship when it was attracted by the magic island, for when it reaches the perimeter of the dimple surrounding a female flower there is a terrific acceleration and the boat slides rapidly downhill and bumps forcefully into its target – a female flower which is still anchored to its mother plant by the long, twisted stalk. The jolt of the collision causes the clump of pollen to be catapulted over to the stigma of the female flower and pollination is achieved. The spiral stalk of the female flower then reverses its action and contracts, pulling the flower beneath the surface, where the young fruit can ripen in safety.

In one of the ribbon weed's close relatives, *Hydrilla lithuanica*, the male flowers go to the surface also. Initially, their calyx lobes, which have the anthers attached to them, are folded outwards and the flowers rest quietly on the water. After a while, however, the anthers suddenly pop up and disperse their pollen over a distance of at least 20 centimetres (8 in). In this fashion, there is a very good chance that it will reach the short papillate stigmas of the female flowers, which rest in a water-dimple similar to ribbon weed. Perhaps it is fair to say that in these plants, and also in *Elodea*, another popular aquarium plant, pollination has become two-dimensional; it takes

place in one plane – the water's surface – and has an element of precision
that is lacking in ordinary wind pollination. It is therefore not surprising
that each pistil has many ovules, and each fruit many seeds. The pollen
grains have a double wall, as they do in terrestrial plants, and plenty of
adhesive. It is hard to escape the conclusion that ribbon weed and its
relatives are the aquatic descendants of plants that were insect-pollinated.

To find the real 'dolphins' among the flowering plants we have to turn to
the so-called sea-grasses, inhabitants of coastal marine habitats in tropical
and temperate regions. They are not at all akin to the land grasses, but count
among their distant relatives the pondweeds and frogbit plants found in
fresh water. All in all there are about 50 species of sea-grass spread over
twelve genera, and in all of them pollination takes place underwater. The
fruits have only one seed each, and the pollen grains have a single rather
124   thin wall. These plants may well have had immediate ancestors that were

Eel-grass, *Zostera marina*, is among the most specialized of the marine hydrophiles. Linear grassy leaves float up from rhizomes rooted in the mud at low-tide level, while inflorescences are borne just below the surface. Cloud-like masses of their worm-shaped pollen grains drift on the tides, and will curl around any narrow object such as the pistil of a female flower.

wind-pollinated. In many cases we find that the pollen 'grains' are worm- or sausage-shaped, even thread-like as in eel-grass (*Zostera*). In *Najas* and *Zannichellia* the grains are spherical or elliptic in the young closed anthers, but they form 'worms' as soon as the anthers open. Their peculiar shape helps to keep them suspended in the water. In *Zostera marina*, the European eel-grass that in the past was often used for bedding material, the pollen is actually lighter than water so that it has a tendency to move to the surface, where it will float. The plants form tiny inflorescences composed of two male flowers topped by a female one. Originally enclosed by a leaf-sheath, they end up close to the surface. The pollen is released in one fell swoop, as a cloud rising to the surface; and some of the sausages may well become entangled in the two thin pistils of a female flower. The inflorescences are protogynous, so that self-pollination is minimized. *Phyllospadix*, another eel-grass with thread-like pollen, is dioecious, which achieves the same end. It is interesting that this habit is found in nine of the twelve genera of sea-grasses that are found in the world. Again, the parallel with wind pollination in the terrestrial plants is obvious. *Ruppia maritima*, a member of the pond-weed family (Potamogetonaceae), probably demonstrates the transition that led from wind- to water-pollination. Each inflorescence is composed of two naked flowers in which two sessile anthers surround four shield-like stigmas. Some flowers open underwater, others above. In the latter case, pollination is taken care of by the wind, with no problems. However, pollen may also fall down on to the surface, where it can pollinate flowers that have opened at exactly this level. The pollen produced by flowers that open underwater, however, has to rise to the surface, where it too has a good chance of pollinating the flowers that opened at that level.

It is interesting to note that abundant rain can also lead to pollination. In the Faeroes, it has been found that pollen floating on the rainwater that accumulates in the flowers of certain buttercups, marsh marigolds (*Caltha*), and bog-asphodel (*Narthecium ossifragum*) can go all the way from the anthers, where it fell, to the stigmas. In one case, that of the common butter-cup (*Ranunculus repens*), it could be demonstrated experimentally that such rain-autogamy does lead to the formation of good fruits. The pollen here must be water-repellent. In other cases, the results are very doubtful. Earlier in this book, we pointed out that for most pollen species exposure to water is fatal. This, perhaps, is the reason why so many botanists take a dim view of the whole idea of rain-pollination!

CHAPTER SEVEN
Pollination and mankind

The story we will present in this our last chapter is, in principle, a heart-warming one – a tale of ancient wisdom and ingenuity, unsung heroes, dramatic new discoveries, and hope for the future. It is especially heart-warming in that it underlines the brotherhood of man and the value of co-operation. At the same time, it may well appeal to those with a liking for irony because it also exposes our fickleness. Human tastes and preferences tend to change in the course of history, and whereas the original aim was to always maximize pollination, simply to get the largest possible crops, we now often strive to prevent it. This applies, for example, to orchids, because pollination here reduces the vase-life of the flowers quite considerably. Many commercial orchid-growers have therefore learned to regard pollinating bees as their mortal enemies. Pollination is, of course, also undesirable in those cases where we want to produce fruits without seeds, as we do with cucumbers. It is fair to say that our knowledge of pollination and fertilization has now become extremely sophisticated, and that we have also learned to appreciate the value of non-sexual reproduction for the creation of desirable new plant types.

The development of agriculture

Thousands of years ago our hunter-gatherer ancestors all over the world scoured the land for edible plant-products; seeds, fruits, young leaves, roots and rhizomes of many kinds. It is quite possible that they sucked the nectar from flowers, just as North American Indians did (and still do) from the nectar-rich, hummingbird-pollinated flowers of Indian paintbrush (*Castilleja angustifolia*), columbine (*Aquilegia formosa*) and *Stachys ciliata*. Indeed the Quinault Indian name for the latter is *gwadjudkolum*, which means 'sweet sucker'. Bushmen and Hottentots in South Africa may have feasted on the nectar of *suikerbossie* or sugar-bush, a *Protea* species.

Throughout the year, all the physical needs of the hunter-gatherers had to be fulfilled by their immediate environment. Their survival depended on mobility, and on a detailed knowledge of the geographical distribution and seasonal growth-patterns of their food plants. However, these hunter-gatherers were no fools. The obvious advantages of saving some of the annual harvest and replanting it at the start of the next growing season were not lost on them. So it is not at all surprising that they gradually exchanged their nomadic life-style for a more sedentary one. The hunter-gatherers thus became farmers, a development that took place independently in various parts of the world, roughly between 15,000 and 5,000 years ago. Rather grandly, we refer to the event as the Neolithic Revolution.

Optimum pollination is critical in the modern commercial production of seed and most fruit crops. The swede rape (*Brassica napus*) is harvested for its seeds, from which a valuable oil is extracted for domestic and industrial uses, the remaining meal also being used in livestock feeds. A full understanding of rape's pollination mechanism is of obvious merit to the farmer and to the plant breeder.

The new life-style provided humans with a better opportunity to exert (not always consciously, perhaps) a certain evolutionary pressure on the food-plants. Primitive man must have selected seeds with good storage characteristics and little or no dormancy. He must also have selected for productivity. Thus, wherever it occurred, the Neolithic Revolution led to the development of staple food crops: maize, potatoes and *Phaseolus* beans in America; millet and rice in the Far East, and various wheats and barleys in the Near East and India. As agricultural techniques improved, food surpluses were created. Populations began to increase and villages gave rise to towns and cities. For the first time, specialist activities and skills could be practised by sections of the community freed from the daily grind of gathering or growing food. Potters, masons, metal-workers, weavers, priests, politicians, administrators, educators and soldiers appeared in society. Some individuals, however, continued to improve their skill in growing crops – and so became specialists in agriculture. They provided the basis for everything else, for even the most advanced and apparently permanent of civilizations is impotent without daily rations. Clearly, man has been practising agriculture with great success now for thousands of years, and for most of this period he remained blissfully unaware of the important role of pollination in crop-production. As we have seen in one of our earlier chapters, it was only in the second half of the eighteenth century that sexuality in the flowering plants was finally recognized. Fortunately, the early lack of knowledge and know-how in pollination matters did not stand in the way of the development of agriculture. A major reason was that most of the crop-plants raised in the early period of civilization were either self-pollinated, such as wheat, barley, oats and beans, or they were pollinated by the wind, like rye and maize. For those plants that depended on insects, there was seldom any shortage of pollinators since damaging pollution was virtually unknown, and the deliberate destruction of insect habitats resulting from human population pressure was still negligible.

A wind-pollinated plant that was exceptional in that it *did* provide problems, at least in certain regions, was the date palm (*Phoenix dactylifera*). The species is dioecious, meaning that there are separate male and female trees. There is strong evidence for the belief that even the ancient Assyrians of Mesopotamia were familiar with the practice of artificial pollination – a famous frieze shows two beings holding male date palm inflorescences over a female date tree.

In general, real pollination problems did not appear until man began to raise plants in areas outside their native habitats. *Ficus sycomorus*, the fig species whose complicated pollination system we have described earlier, must have been brought over to Egypt without its native pollinator, *Ceratosolen arabicus* (or perhaps the pollinator could not tolerate the Egyptian climate and so died out). Still, the Egyptian trees produce mature, good-tasting figs – a common feature of Egyptian markets. Upon inspection, one finds that every single one of these figs bears a fairly deep, circular or elliptical incision. The fig-growers inflict the wounds at an early stage in the life of the fruit, with the aid of special metal hooks. (These implements have also been found in ancient Egyptian tombs, indicating that the method goes back very far in time.) As it does in many plant tissues, the wounding produces ethylene, the gas that is known to act as a fruit-ripening hormone.

This relief found by Layard at Nimrud in Mesopotamia dates back to about 1500 B.C. providing us with evidence that the practice of artificial pollination is certainly not a modern one. It depicts two divine winged creatures each holding a male date palm inflorescence over a female tree.

Ethylene triggers the pollination events and stimulates cellular respiration in normal figs that contain the pollinating fig-wasps. It is responsible for the softening of the fruit wall that makes possible tunnel-digging by the male wasps and subsequent escape of the females.

The more complicated a pollination system, the greater the chance that something will go wrong. It is therefore not at all surprising that the 'classical' Mediterranean fig (*Ficus carica*) also presented problems. In this case, the trees that produce fruits for human consumption have only long-styled flowers; the short-styled ones suitable to serve as receptacles for a gall wasp's eggs, and thus create a new generation of pollinators, are entirely lacking. The fig-grower must therefore practise 'caprification', which means that he must suspend branches of the wild 'caprifig' (which does produce pollinators) in the crowns of his orchard trees. Unfortunately, the principle of caprification was overlooked when, for commercial purposes, *Ficus carica* was introduced into California. Initially, no fruits were produced at all, and this situation was not remedied until the pollinating gall wasps were also introduced to the area.

As to the introduction of plant species into other continents, an event comparable to a quantum-leap in its dramatic effects was the discovery of the New World, and especially the conquest of Mexico by Cortés in 1519. Both agriculture and horticulture had reached a high level of development under the Aztecs.

The Central American vanilla plant presents us with a baffling puzzle. The commercial product represents the slender fruits (the 'vanilla pods') of vanilla orchids, mostly of *Vanilla planifolia*. Vanilla, one of the Aztec products valued most highly by the conquering Spaniards, is now grown in many tropical countries, including Madagascar, Java and Polynesia. But no matter where it now occurs (and this includes its native Mexico), it is always pollinated by hand. A natural pollinator has never been found. The claims made for honey-bees in Puerto Rico can safely be discarded, since these pollinators are not native to America. It is difficult to escape the conclusion

that for the past several hundred, or even thousand years vanilla orchids have always been pollinated by hand. This would indicate a most remarkable degree of sophistication on the part of the Aztecs, since pollination here involves very delicate manipulation of the flower parts.

Enlisting the help of bees

Vanilla is an extreme example of a so-called oligophilic plant, one which depends for its pollination on a single pollinator species or on a small group of different but related pollinators. It goes without saying that the members of this plant group run a particularly high risk of encountering pollination difficulties when they are transferred from their native habitat to some faraway country. Red clover (*Trifolium pratense*) is a well-known example. The nectar-rich flowers are self-incompatible and depend for their pollination (and, consequently, seed-production) on long-tongued bumblebees such as *Bombus hortorum*. The reason is that the flower-tube is so long (9–10 millimetres [$\frac{1}{3}$–$\frac{2}{5}$in]) that honey-bees fail miserably here; their average tongue-length is only 6 millimetres ($\frac{1}{4}$in), so they cannot get at the nectar in the base of the tube. When red clover was introduced to countries where it did not previously grow, such as New Zealand, seed set was initially very low. After introduction of suitable European bumblebee species, it improved dramatically. However, using bumblebees has its disadvantages too, one of the main problems being that of building up sufficiently high populations of these useful creatures. Other solutions to the clover problem have therefore been tried, including attempts to breed honey-bees with longer tongues, and

Vanilla-growers rely entirely on hand-pollination. This orchid, *Vanilla planifolia*, the fruits of which yield vanilla essence, bears racemes of 20 or more flowers which open in succession. Each flower lasts but one day. The 'pollinator' uses a pointed stick to mutilate the flowers so that the pollinia and stigma can be pressed together between thumb and forefinger. The entire plantation must be hand-pollinated every morning to ensure maximum seed-set. A trained operator can deal with 2,000 flowers a day.

clover races that are either self-compatible or have shorter flowers. In many countries, these breeding programmes have been so successful that bumble-bees are no longer needed.

In the United States, both the honey-bee and alfalfa (*Medicago sativa*), an extremely important crop plant, can be regarded as aliens. However, they did not come from the same part of the world, and so do not form a natural pollination pair. (Alfalfa is a plant of Near Eastern or Mediterranean origin, while honey-bees got their start as a species in tropical Asia.) So, it is not at all surprising that there have been serious problems with alfalfa pollination. Alfalfa flowers have an explosive mechanism very similar to that of Scotch broom. When a not-too-small insect visitor lands on an alfalfa flower, its weight unlatches the mechanism that allows the stamens to snap upwards in true jack-in-the-box fashion and powder the visitor's underside with a cloud of pollen grains. The process is called 'tripping'. Honey-bees, mainly interested in alfalfa flowers for their nectar, soon develop an aversion to-wards the explosions. They learn to steal the nectar through a natural slit, and from then on will trip the flowers only by accident – in practice, in about one case in 20. As pollinators they are therefore not really worth their salt. Fortunately for alfalfa growers, bumblebees and various solitary bees readily visit the flowers. In America, the two most important species are the alkali bee (*Nomia melanderi*) and a leafcutter bee (*Megachile rotundata*). On alfalfa, these animals collect both nectar and pollen (the ingredients for the bee-bread on which their larvae thrive), but they collect mainly pollen. They also trip the flowers efficiently, and twice as rapidly as honey-bees.

The problem with solitary bees is that their numbers are unpredictable, and usually too low to reliably pollinate large acreages: for instance, 5,000 nesting female leafcutters are needed to pollinate each hectare of alfalfa. Consequently, the management of alkali and leafcutter bees has become big business. Alkali bees are provided with large beds of soil of just the right composition and moisture content for digging their burrows. In such mani-cured surroundings, the concentration of nesting burrows can be increased to between 2,000 and 3,000 per square metre compared with 500 or so in natural soil banks. Leafcutters, which naturally nest in small holes in wood or stone, or in narrow hollow stems such as those of dahlias and bamboo, are provided with suitable 'housing-developments' in the form of wooden or plastic blocks with a large number of holes drilled in them.

Such housing-developments, filled with leafcutter brood, can be sent from place to place to be put in alfalfa fields as needed. In the Pacific Northwest, such management practices have been so successful that the yields of alfalfa seeds have been boosted to nearly four times the national average. In some places, one can even see road signs urging motorists to drive carefully because alkali or leafcutter bees are crossing.

Bumblebees (*Bombus* species) also do a good job on alfalfa, cotton (*Gos-sypium*) and broad beans (*Vicia faba*), and are particularly effective on the various *Vaccinium* species referred to as blueberries, huckleberries and cran-berries. The crucial role of bumblebees in huckleberry pollination was dramatically demonstrated by the eruption of Mount Saint Helens in Wash-ington, US, in 1980. In a wide area stretching out from the mountain in an east-northeasterly direction, the ash rains resulting from the catastrophe wiped out bumblebees completely. Many huckleberry plants survived the

131

eruption: in fact they benefited from it because the ash improved the nutritional value of the soil. However, the next harvest season there were hardly any berries.

In contrast to honey-bees, bumblebees are well adapted to cold climates and are active in such places as Alaska and the high Alps. They work longer hours than honey-bees and visit twice as many flowers per foraging trip. Unfortunately, to domesticate them effectively is not an easy task; they do not take happily to artificial housing and do not like being moved from place to place. Also, bumblebee colonies rarely muster more than 400 workers, compared with the honey-bee's 40,000 to 60,000. There is hope for the future; but for the time being the best policy we can follow if we want to boost pollination by bumblebees is to leave plenty of waste ground, well stocked with small rodents in whose tunnels and abandoned nests bumblebees can set up colonies.

On the big island of Honshu in Japan, a mason bee (*Osmia cornifrons*) has been used successfully since 1958 for the pollination of apple trees. This bee will nest in thin bamboo stems and hollow reeds which the farmers place on shaded platforms in or near the apple orchards. Very conveniently, the bees begin to fly about two weeks before apple blossom time. The great advantage they have over honey-bees is that they remain active at temperatures as low as 7°C (about 45°F) – well below that at which honey-bees fly. In the central part of Washington State in the United States, a prime apple-growing region, *Osmia lignaria* has also been found willing to accept man-made dwellings – and its use has again vastly improved crop yields. The European *Osmia rufa* can also be induced to nest in drinking straws. This species often develops a preference for *Rubus* species, and so holds out promise for the pollination of raspberry and blackberry crops.

The sunflower (*Helianthus annuus*), a member of the Compositae that originated in Central America, is a crop of increasing economic importance. The seeds yield edible oils, animal feedstuffs and health-food products for human consumption. Each sunflower inflorescence may contain up to 2,000 florets, which are visited for their nectar by various insects, the most commercially important of which are honey-bees. Many growers optimize their yields by introducing hives among the crop.

A far from complete picture

While pollinator management is of immediate and widespread use in many circumstances, a surprising amount of ignorance still surrounds the whole subject. There are still many crops whose pollinators are not known with any degree of certainty, and a good many that probably need no animal assistance at all.

One of the many 'mystery' crops is cacao (*Theobroma cacao*), which bears its flowers directly on its branches and main stem. The flowers have no nectar, and very little scent. Various animals have been proposed as the plant's pollinator – including thrips, ants and flies – but the theory currently favoured is that pollination is effected by female *Forcipomyia* flies visiting the plants in order to probe the tissues for sap.

The various citrus fruits – oranges, limes, lemons, grapefruits and tangerines – bear fragrant flowers with abundant nectar and are consequently visited by a host of insects of all shapes and sizes. And yet most of these widespread and commercially important fruit-plants are self-pollinated and can fruit perfectly well without flower visitors. Some, however, such as clementines, are self-incompatible and must have assisted cross-pollination.

Although grapes (*Vitis* spp) have been intimately associated with human history for the last 6,000 years, surprisingly little is known of even their exact pollination requirements. It was realized that while all the flowers of cultivated vines are hermaphroditic, they are of two types – some basically male, the others female – and added to this is the fact that some varieties are self-fertile but not automatically self-pollinating, while others need pollen from another variety for fruit-set to occur. In this latter category falls the Spanish 'Almeria' grape which, when introduced to California, was a great disappointment. After many years of abysmal harvests, the growers realized their mistake: by some oversight they had failed to introduce the necessary pollinizer varieties at the same time.

Such is the state of our knowledge; in some areas very full and able to support sophisticated plant breeding and commercial exploitation, in others inadequate to say the least, leaving farmers with little more than tradition to support their hit-and-miss activities.

The role of the plant physiologist

If we try to find a common denominator in the pollination studies we have so far described, it will be readily apparent that they were all based on the manipulation of pollinators. From the results we can conclude that there is nothing wrong with this approach, and we can look forward to further valuable contributions by entomologists and agriculturists. However, a necessary counterpart is the manipulation of plants, or rather of plant behaviour and plant properties, and for this reason plant physiologists have now become great allies of entomologists. Later on we shall see that geneticists too have a crucial role to play in the development of applied floral biology. The following examples will help to explain the plant physiologist's work.

Regulation of flowering. It is now generally recognized that in many plant species flowering is controlled by day-length. In so-called short-day plants, exemplified by cocklebur (*Xanthium pennsylvanicum*), a certain critical

133

length of the night-time period of darkness has to be exceeded if blooming is desired. For most of the cocklebur types found in North America, the critical period is about eight and a half hours. Its interruption by a brief flash of light (preferably red) prevents flowering. Hawaiian sugar cane happens also to be a short-day (or long-night) plant. Flowering is undesirable here because it is a prerequisite of pollination, and the development of fruits and seeds to which pollination would lead is such a heavy drain on the system of the sugar cane plant that it would lower the sugar level, which is obviously unacceptable. In the autumn, the decrease in day-length that normally induces flowering in this sugar cane can be counteracted by artificially lengthening the day with the aid of electric lights. This, however, is expensive, and a better solution is simply to give the plants a brief 'daily' illumination some time after darkness has set in.

Regulation of sex ratios. In many plant species, sex-expression is very variable. In pumpkin and other members of the cucumber family, for instance, the application of ethylene gas (or of an ethylene precursor such as Ethrel) tends to induce femaleness in the early flowers. In the course of the flowering season, the shoots of pumpkin vines normally produce male flowers first, followed then by hermaphroditic and finally by female ones. The pumpkins that develop from the latter often come so late in the autumn that they run a real risk of being destroyed by frost. For the pumpkin-grower it is therefore advantageous to have the female flowers appear as early in the season as possible, and this can be achieved by spraying with Ethrel.

Fruits formed parthenogenetically. Plant physiologists have also demonstrated that a flower's ovary can develop into a practically normal fruit even if it does not contain viable seeds. That is, pollination and fertilization are not always necessary for fruit formation. In certain orchids the pollinia are rich in indole acetic acid (auxin), and it is this compound that does the trick. Fruit development follows when the pollinia are placed on the stigma, even after they have been killed by a heat treatment. Good, edible, *seedless* tomatoes and cucumbers can likewise be produced by treatment with certain plant hormones. It is also worth mentioning that there are some varieties which spontaneously produce parthenogenetic fruits. In certain cucumber-producing regions in the Netherlands, these facts have led to interesting changes in people's attitudes towards bees. Plants in the cucumber family depend on these animals for normal pollination, followed, naturally, by the formation of seed-containing fruits. Consequently the bees were very highly regarded. However, with the advent of seedless cucumbers – vastly preferred by housewives for their 'neatness' and therefore commanding a higher price – the bees suddenly became *persona non grata*.

The geneticist's contribution: plant breeding

It has been estimated that as early as 3000 B.C. all the broad categories of food-plant currently under cultivation were already being utilized. As we have seen, the methods employed by early man to arrive at this state of affairs were amazingly simple. Essentially, he simply took what was already there. Plant individuals that had something to offer were collected and grown, and if this was successful an endless series of offspring generations was produced. There was very little change, though once in a great while

134

an individual of a new type might show up in a population of crop plants as the result of sudden and spontaneous change in the plant's genetic make-up.

When offspring are produced sexually, it is common to find them differing from one another, and from their parents, in a number of ways. Another cause of genetic variation between sexually produced offspring is gene mutation. Occasionally an individual is produced carrying a completely new gene, which may have a noticeable effect on some facet of its biology. While most mutations are lethal, some are neutral and a few beneficial in their effect. For instance, the genetic basis of the range of colours and patterns now found in cultivated poppies (*Papaver rhoeas*), compared with the standard red petal with central black blotch of the wild type, is to be found in two particular mutations, one giving a white edge to the flower and another altering the colour of the centre. These mutations were spotted by poppy-fanciers in the later 1800s and the individuals carrying them were used for controlled cross-breeding experiments, the results of which are today's multi-coloured varieties.

The rate at which mutations occur in the wild is in the order of one in every 10,000 to 100,000 gametes. What is more, most naturally occurring mutations are recurrent, and lethal in their effects. In order to speed up the mutation rate, plant breeders have taken to subjecting plants to a variety of mutagens – mustard gas, gamma-rays, X-rays, neutrons, alpha particles – largely without significant gains. There have been exceptions, however, and improved varieties of barley, rice and peanut have appeared as a result of such artificially induced mutation.

The vast majority of plant breeding to date has been based on the simple but time-consuming technique of carrying out controlled, or random, cross-pollination experiments, followed by the selection of the best individuals from among the progeny. Unfortunately, the results of mutation and gene recombination are unpredictable, and although nearly all major crop and garden plants have been improved by professional and amateur plant breeders, these advances have either been due to pure luck or are the results of colossal enterprise. For example, over 20 million potato seedlings have been raised in the last 50 years in the continuing effort to find better varieties – largely without success.

Progress is also slowed down because in many cases it takes such a long time for the results of a plant breeding programme to reach a stage at which they can be judged. 'Golden Delicious', the ubiquitous supermarket apple, originated in the United States from a self-sown seedling spotted in 1890 in Clay County, West Virginia. The pip from which the seedling grew came from a Grimes Golden apple: its 'father' was unknown. The young plant was taken to a local nurseryman, who developed the promising newcomer – but it was not until 1914, nearly a quarter of a century later, that this exceptionally successful variety was released on to the market.

The breeding of roses offers another example of the enormous efforts undertaken by commercial breeders. There are about 250 species of the genus *Rosa*, and many thousands of cultivated varieties. The majority of the latter are the result of natural variation following the crossing of two parental stocks, and in the case of the rose this activity has been going on for hundreds of years. Roses have been bred in the Far East for as long as they have in Europe, and by the eighteenth century there already existed a large number of cultivated varieties. With the burgeoning of trade in the eighteenth and nineteenth centuries many oriental varieties found their way to Europe and so enriched the available breeding stock. European flowers tended to have a short flowering season, while the oriental growers had selected varieties that would bloom continuously for weeks. The mixing of the two groups brought about huge numbers of new and popular types.

Today, such is the scale of the rose-growing industry that several million rose seeds are germinated every year in the quest for new varieties with commercial potential. It takes at least eight years of careful management, and ruthless selection, to produce a new variety. The crosses are made in May or June, the hips are ripened and the seeds are removed and vernalized ready for planting in February of the next year. Between 10 and 50 per cent of them fail to germinate. Those that do germinate produce flowers in about four months, although they are still only seedlings about 15 centimetres (6 in) tall. This is the stage at which the breeder can make his first selection. He usually finds that between 10 and 40 per cent are worth retaining for further trials, the rest being consigned to the incinerator. The 'chosen few' are budded out on to field-grown rootstocks that summer (perhaps five new plants from each seedling). The buds remain dormant until the next year (year three in the progression), when the breeder will have a further chance to assess his new lines. Again, the majority will be rejected. Year four sees another multiplication via budding, and a further weeding out of all but the most promising plants. The varieties that survive to year five are the élite; perhaps one out of every 1,000 seedlings raised in year two. During years five and six the new roses are further multiplied by budding and grafting, and are distributed among rose growers and national organizations for further trials. By year six a decision will have been made as to which varieties (if any) are suitable for commercial exploitation. Years six to eight are taken up with vegetative multiplication – building up sufficient stocks for the 'débutante's' launch on to the market.

In the huge family of the orchids we find numerous exceptions to the 'rule' that one cannot hybridize different plant species. Crosses have even been obtained between orchids so different that according to all normal criteria of colour, size, shape and flower pattern they belong to completely different genera. In such cases there are obviously no chemical incompatibility barriers capable of preventing the hybridization. Perhaps there has been no selection for such barriers in the process of evolution, because, as we have seen in *Coryanthes* and *Ophrys*, the pollination mechanisms of orchids are usually so outlandish and specific that the danger of hybridization in nature is virtually non-existent. Consequently there are now thousands upon thousands of commercial orchid hybrids. In fact, most of the flowers sold by florists come from such plants.

Most orchids produce an enormous number of seeds per fruit. One vanilla pod, for instance, may contain more than a million. The wind-dispersed seeds are like tiny specks of dust – far too small to contain enough reserve material for a developing seedling to draw on. When a seed has been deposited in a favourable spot – usually on the outside of a tree branch – it may take up water and go through a limited development that leads to the formation of a 'protocorm' – a rather undifferentiated little body. No further development will take place unless the protocorm is invaded by a fungus.

In the commercial breeding of orchids, such as *Cymbidium*, hand-pollination is adopted to ensure the desired result – other pollinators are excluded. The operator removes a pollen ball from one flower (A) using a sharpened stick. A 'cement', which in nature would glue the pollen ball to an insect's back, here serves to hold it on the tip of the stick. The operator must then peel off and discard the cap-like casing (B) before relocating the now exposed pollen on to the receptive stigma of another flower (C).

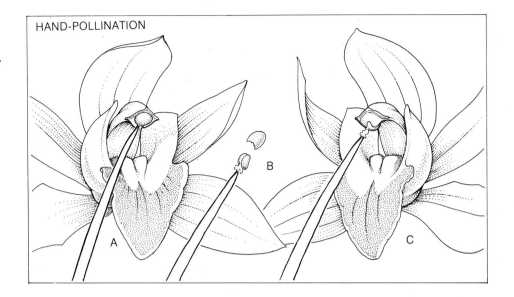

HAND-POLLINATION

In an ideal situation, a certain equilibrium is established between the two organisms which leads to a permanent symbiosis, or mutually beneficial relationship. The fungus probably provides the protocorm with nutrients and vitamins, or perhaps facilitates the uptake of nutrients from the environment. The adult orchid still contains, at least in the peripheral parts of its roots, fungal mycelium.

Clearly, raising orchids in the laboratory from seeds is no mean undertaking. The sterilized seeds, kept on an agar medium, have to be inoculated with fungus mycelium taken from a pure culture. Development is then allowed to take place under sterile conditions. It was a pleasant surprise when the American botanist Knudson discovered, some 50 years ago, that a number of orchid species could develop successfully without fungus by keeping them on a special medium designed by him. His method is now being used on a large scale by most commercial orchid-growers. However, it remains labour-intensive and thus expensive, and does not abolish the long waiting period between seed germination and flowering. This is the reason why Knudson's method is now being replaced, where possible, by the very modern one of 'meristeming' – a vegetative method based on the culture and subsequent development into plants of pieces of meristem, that is, of an embryonic tissue found in the vegetation tips. Rapid multiplication of superior orchid types has thus become possible. The 'offspring' plants are all very, very similar to each other; so much so that they form a clone. The commercial possibilities of the method can hardly be overestimated.

Pushing back the frontiers

Given the enormous expense in time, effort and money involved in most plant breeding programmes, it is hardly surprising that in recent years breeders have been searching (successfully, as we shall see) for methods that are both time-saving and more predictable. One such method is androgenesis, the production of whole plants from pollen grains.

Since scientists achieved a breakthrough in culturing techniques in the middle sixties, it has been possible to induce pollen grains to form multi-

137

cellular structures that eventually differentiate into entire haploid plants. This androgenesis, as it is called, has now been found in more than 250 species from 25 plant families, and the new techniques have yielded improved cultivars of tobacco, rice, wheat, maize, rape, rye, sugar cane and rubber. Since the plants are haploids, the effects of recessive genes and mutations manifest themselves very clearly and such unpromising lines can be rejected.

If varieties are discovered which do show promise, homozygous pure lines can be produced in one further step by doubling the chromosome number – for example by using colchicine, a compound found in extracts made from the autumn crocus (*Colchicum*). Androgenesis gives an enormous saving in time and resources when compared with the conventional techniques of sexual breeding.

Epilogue: The best is yet to come (but not soon!)

We all know that speculating about the future is risky business. However, at this particular juncture in history, let us therefore look into our crystal ball. What do we see? Frankly, a crazy quilt of positive and negative factors.

On the one hand, we can look forward to further spectacular triumphs in the areas of applied pollination and plant propagation, based both on time-honoured and brand-new approaches. For example, the pollination of the West African oil palm (*Elaeis guineensis*), one of the world's most important crop plants, was elucidated only four or five years ago. It has now been established beyond any reasonable doubt that the natural pollinator

With starvation facing a large proportion of the world's population, the breeding of higher-yielding, hardier, disease-resistant crop-plants, especially cereals such as wheat (*Triticum aestivum*), is a major concern.

is a weevil, an *Elaeiodobius* species. While observing all the necessary precautions, biologists have recently introduced it into other oil-palm-growing regions such as Malaysia, where the yield left a great deal to be desired. The result has been a phenomenal yearly gain in oil production. On the other hand, one can legitimately ask what the ultimate value of such short-range successes is. The culture of African oil palms in Brazil, where these trees have been introduced, has led to terrible erosion problems and thus to the ecological rape that is going on in that part of the world.

We can acknowledge that the techniques of androgenesis, protoplast culture, artificially induced mutation and genetic engineering, together with the classic approach of cross-pollination and selection, offer a vast new potential to the plant geneticist. It is now quite conceivable that one day our descendants will be able to assemble new plant types 'to order' by splicing together combinations of genes from many species. At the same time, however, we are destroying whole populations and communities of plants, especially in the tropics. Numerous species, genera, perhaps even families, are threatened with extinction or have already been wiped out. Many have never been named, let alone studied, but one thing is certain: the loss of genetic material is incalculable. Our pitifully slow creation of new genetic types can never compensate for the wanton, wholesale destruction of the vast natural reservoir of genetic material. In such a situation, it is easy to yield to despair. However, we do not wish to end this book on a sombre note. Admitting that things will undoubtedly get worse before they get better, we can offer at least a small ray of hope. People are beginning to realize that for the very survival of mankind a change in attitude is imperative. Reckless population growth has to be curbed. Furthermore, there is nothing wrong in adopting the attitude of the North American Indian, who lived in harmony with Nature rather than exploiting it ruthlessly. There are at least a few favourable signs. In the Netherlands, there has been a popular 'rebellion' against the incessant spraying of roadsides with chemicals that has led to the demise of colourful wildflowers, 'weeds' and beneficial insects and their replacement by a deadening sameness of green grass. With very few exceptions, roadside spraying is now forbidden by law, and the result has been a return of colour that gladdens the heart.

It is obvious that we need education and eye-openers, and without any preachiness, pretence, smugness or condescension we venture to express the hope that in its own humble way our present book can indeed be an eye-opener. The world we live in is still a beautiful place. Fate has been kind to us – the writers of this book – in that it has given us a better chance than most people get to uncover Nature's treasures. We are grateful, and offer you this book in a spirit of sharing.

GLOSSARY

ANDROGENESIS: the production of a whole plant from a pollen grain.

ANEMOPHILY: wind-pollination.

ANGIOSPERM: one of the flowering plants, a major group of seed plants in which the seeds are borne and develop within a closed ovary.

ANTHERIDIUM: male sex organ of ferns releasing sperm cells (also of algae, fungi, liverworts and mosses).

ANTHOCYANINS: a group of water-soluble pigments found in vacuoles of flower, stem or leaf tissues, responsible for many red, purple and blue colours according to the particular chemical nature of the compound present and to the acidity of its solution.

ANTHOPHILOUS: flower-loving; applied to animals that can act as pollinators.

ANTHOXANTHIN: a yellow pigment in flower tissues.

ARCHEGONIUM: female sex organ of ferns containing the egg cell (also of liverworts, mosses and most gymnosperms).

AUTOGAMY: self-fertilization.

BETACYANIN: a nitrogen-containing red or purple pigment that replaces anthocyanin in most of the Centrospermae, e.g. in beets (*Beta*) and *Bougainvillea*.

BILATERAL SYMMETRY: having two identical halves; symmetrical about one plane only.

CALYX: the outermost whorl of floral parts: sepals.

CAROTENE: an orange-yellow pigment in chloroplasts, sometimes found in flowers, fruits and roots in the absence of chlorophyll (e.g. gives orange colour to carrot roots).

CARPELLATE: being entirely female, having pistils or carpels, but no stamens.

CARPEL: the female reproductive part of a flower which, alone or in combination with other carpels, forms the pistil composed of ovary, style and stigma.

CATKIN: a special type of inflorescence usually consisting of single-sex flowers around a common stalk.

CHASMOGAMOUS: having flowers which open to be pollinated by wind or by insects etc (opposite to cleistogamous).

CHIROPTEROPHILY: pollination by bats.

CHLOROPHYLL: the pigment present in the chloroplasts, giving the green colour to all algae and most higher plants if not overridden by other (red, blue or brown) pigments; instrumental in the synthesis of food materials from carbon dioxide and water with the aid of light-energy.

CHLOROPLAST: a small granule within the cytoplasm of a plant cell containing chlorophyll and other compounds active in photosynthesis.

CLEISTOGAMOUS: having flowers (often inconspicuous) which do not open and are self-fertilized.

CLONE: a group of individual plants, derived by vegetative means, all with a common ancestor.

COLUMN: the central part of an orchid flower bearing united anther and stigma.

COMPOSITE: a member of the Compositae family of flowering plants, characterized by having flowers in a capitulum (a complex inflorescence with a central disc of petalless hermaphrodite florets surrounded by a ring of female or sterile single-petalled ray florets, the whole functioning as a single flower).

COROLLA: the whole whorl of petals.

CROSS-POLLINATION: pollination in which pollen is transferred from a particular flower to another found on a different plant of the same species.

DICHOGAMY: a floral condition in which male and female parts mature at different times, preventing self-pollination.

DIOECIOUS: having entirely male flowers on one plant and female flowers on another.

DIPLOID: having a double set of chromosomes, i.e. having twice the number that a sex cell has.

DISTYLY: a genetically controlled condition in certain plant species causing some individuals to have flowers with long styles and short stamens while other individuals have flowers where the reverse is true. Promotes cross-pollination (outbreeding).

ELAIOPHORES: oil-producing parts of a flower or inflorescence.

ENDOSPERM: nutritive tissue surrounding the embryo within a seed, or adjacent to it.

ENTOMOLOGY: the study of insects.

EPHYDROPHILES: aquatic plants bearing flowers at the water surface or sending them there, generally pollinated by wind.

EPIPHYTE: a plant individual which grows attached to another plant, but using it only for support, not for nutrition.

EUGLOSSINE BEES (syn: orchid bees): neotropical bees belonging to the tribe Euglossini (Apidae), characterized by metallic or bright body-coloration, by hind legs modified into a pouch in the males, and by a long tongue which enables these bees to forage in flowers with long spurs or deep corollas.

GAMETE: a sex cell; haploid.

GAMETOPHYTE: the gamete-producing phase of species displaying alternation of generations; composed of haploid cells only.

GYMNOSPERM: member of a group of plants of which the ovules (and, later, the seeds) are exposed rather than being borne within a closed ovary.

HAPLOID: having a single set of chromosomes. Refers to a sex or germ cell.

HELIOTROPIC: sun-tracking; applied to flowers that can rotate so that they are always facing towards the sun.

HEMIPTEROUS: refers to insects within the order Hemiptera possessing two pairs of wings, and mouthparts adapted for piercing and sucking.

HETEROSTYLY: a genetically controlled condition in certain plant species in which the stamens and styles come in two or three well-defined size-classes (short and long, or short, middle-sized and long), in such a way that a flower with a style of a certain length cannot have stamens of that same length, while furthermore different plant individuals each have flowers of one type only. Promotes cross-pollination (outbreeding).

HOMOZYGOUS: refers to a condition in which the genes in a particular 'locus' (spot) on a pair of homologous chromosomes are alike.

HYPHYDROPHILE: an aquatic plant with submerged flowers, being pollinated entirely under water.

INFLORESCENCE: an assembly of flowers acting as a higher-order unit in pollination.

LABELLUM: the lower petal of certain flowers, particularly in orchids and members of the mint family (Labiatae); often formed into a lip.

MEGAGAMETOPHYTE: a female gametophyte formed from a megaspore.

MEGASPORANGIUM: a sporangium producing megaspores.

MEGASPORE: in plant species producing two kinds of spores, the spore-type that gives rise to the female gametophyte.

MERISTEM: a group of undifferentiated, 'embryonic' cells found at a growing point on a plant, capable of giving rise to any tissue type.

MERISTEMING: a process leading to the culture of whole plants from excised meristem cells.

MICROGAMETOPHYTE: a male gametophyte formed from a microspore.

MICROSPORANGIUM: a sporangium liberating microspores.

MICROSPORE: in plant species producing two kinds of spores, the spore-type that gives rise to the male gametophyte.

MONOECIOUS: having separate male and female flowers, but both borne on the same plant.

MUTAGEN: an agent capable of inducing genetic mutation.

MYCELIUM: the vegetative part of a fungus.

MYCORRHIZA: a mutually beneficial relationship usually between the mycelium of a fungus and the roots of a higher plant.

NECTARY: a nectar-secreting gland.

NOTOTRIBIC POLLINATION: a mechanism by which the anthers and stigma of a flower touch insect visitors' backs and thereby facilitate successful pollination.

OLIGOPHILIC: dependent on a small group of different but related pollinators.

ORNITHOPHILOUS: pollinated by birds.

OSTIOLUM: the small mouth-like opening found in figs.

OVULE: the structure which develops into a seed after fertilization of the egg-cell inside it.

PAPILLAE: elongated protuberances on the surface of cells or tissues; can be multicellular (adj. papillate). Responsible for the 'velvet' effect of certain flowers, and for the rough tongues of certain bats, for example.

PARTHENOGENESIS: production of new individuals from unfertilized egg cells.

PERIANTH: the floral envelope as a whole (calyx and/or corolla).

PISTIL: the female organ of a flower, consisting when complete of ovary, style and stigma.

POLLINARIUM: a detachable structure found in certain plant species including orchids, consisting of a stalked pollinium with attached adhesive disc which facilitates transference by an animal carrier to another flower.

POLLINIUM: a pollen mass composed of large numbers of cohering pollen grains.

PROTANDRY: a condition in a flower in which the stamens reach maturity and shed pollen before the female organs in that same flower are receptive, hence preventing self-fertilization from occuring.

PROTOCORM: in orchids, the first and still undifferentiated structure formed by a germinating seed; in nature it requires invasion by a fungus before it will develop into a normal seedling.

PROTOGYNY: the condition in a flower in which the pistils reach maturity and become receptive before the stamens in that same flower are ripe and start shedding pollen, hence preventing self-fertilization.

PROTOPLAST: a nucleus with its cytoplasm; 'a cell without its cell wall'. Protoplast culture is a technique by which whole plants are generated from such material.

PSEUDO-COPULATION: in botany, the attempted copulation between the males of certain wasp and bee species and the flowers of certain orchids that mimic their females.

ROSTELLUM: a beak-like projection from the stigmatic surface or column of an orchid flower.

SELF-FERTILE: the female gametes being receptive to fertilization by male gametes from the same flower.

SELF-POLLINATION ('selfing'): the transference of pollen grains from the anthers to the receptive stigma of the same flower or of other flowers on the same plant.

SPADIX: the fleshy central axis of certain inflorescences, such as those of arum lilies, bearing the small flowers or florets.

SPATHE: a large bract enclosing a flower cluster, or the flower-bearing spadix of arum lilies.

SPORANGIUM: a spore-producing organ.

SPOROPHYTE: the spore-producing phase of a species displaying alternation of generations.

STAMINATE: being entirely male, bearing only stamens.

STERNOTRIBIC POLLINATION: a mechanism by which the anthers and stigma of a flower touch the underside of insect visitors' bodies and thereby facilitate pollination.

SYNSTIGMA: a felty mat-like surface composed of interwoven, often hairy, stigmas belonging to crowded female flowers, e.g. within the inflorescence of a fig.

SYRPHIDS (Syrphidae): hoverflies and droneflies; two-winged insects, the former often resembling yellowjacket wasps with an ability to hover in mid-air, the latter resembling honeybees; most are pollen and nectar feeders, but a few are nectar specialists.

VERNALIZATION: a seed treatment involving long-lasting exposure to cold and moisture, or sometimes to a period of light, to promote more rapid development of the subsequently germinated plant, especially with respect to earliness of flowering.

XANTHOPHYLL: a yellow or orange pigment usually associated with chlorophyll within the chloroplasts of green plants.

ZOOPHILY: animal-pollination.

ZYGOTE: the fusion product of a pair of gametes or sex cells; diploid.

FURTHER READING

GENERAL

Armstrong, J. A., Powell, J. M., and Richards, A. J. (Eds.) (1982). *Pollination and Evolution*. Roy. Bot. Gard., Sydney.

Barth, F. G. (1982). *Biologie einer Begegnung*. Die Partnerschaft der Insekten und Blumen. Deutsche Verlags-Anstalt, Stuttgart.

Bell, G. (1982). *The Masterpiece of Nature: the Evolution and Genetics of Sexuality*. Univ. Calif. Press, Berkeley, Cal.

Bristow, A. (1978). *The Sex Life of Plants*. Holt, Rinehart and Winston, New York.

Corner, E. J. H. (1964). *The Life of Plants*. World Publ. Co., Cleveland.

Faegri, K., and van der Pijl, L. (1979). *The Principles of Pollination Ecology*, 3rd edn. Pergamon Press, Elmsford, N.Y.

Futuyma, D. J. (1979). *Evolutionary Biology*. Sinauer; Sunderland, Mass.

Holm, E. (1979). *The Biology of Flowers*. Penguin Books, Harmondsworth, Middlesex.

Jaeger, P. (1961). *The Wonderful Life of Flowers*. Dutton, New York.

Jones, C. E., and Little, R. J. (Eds.) (1983). *Handbook of Experimental Pollination Biology*. Van Nostrand Reinhold, New York.

Kerner von Marilaun, A., and Oliver, F. W. (1904). *The Natural History of Plants*. Gresham, London.

Klein, R. M. (1979). *The Green World: an Introduction to Plants and People*. Harper & Row, New York.

Knoll, F. (1956). *Die Biologie der Blüte*. Springer, Berlin.

Kugler, H. (1970). *Einführung in die Blütenökologie*, 2nd edn. G. Fischer, Stuttgart.

Meeuse, B. J. D. (1961). *The Story of Pollination*. Ronald Press, New York.

— (1974). 'Pollination'. *Encyclopaedia Britannica*.

— (1984). *Pollination*. Oxford Carolina Reader.

Müller H. (1881). *Alpenblumen, ihre Befruchtung durch Insekten und ihre Anpassungen an dieselben*. W. Engelmann, Leipzig.

— (1883). *The Fertilisation of Flowers* (trans. D'A. W. Thompson). Macmillan, London.

Pelt, J. M. (1970). *Evolution et Sexualité des Plantes*. Horizons de France, Paris.

Proctor, M., and Yeo, P. (1973). *The Pollination of Flowers*. Collins, London.

Percival, M. S. (1965). *Floral Biology*. Pergamon Press, Oxford.

Raven, P. H., Evert, R. F., and Curtis, H. (1976). *Biology of Plants*, 2nd edn. Worth Publ., New York.

Real, L. (Ed.) (1983). *Pollination Biology*. Academic Press, New York.

Richards, A. J. (Ed.) (1978). *The Pollination of Flowers by Insects*. Linnean Soc. Symp., No. 6. Academic Press, London.

Ruscoe, Q. W. (Ed.) (1979). 'Reproduction in Flowering Plants'. *New Zealand J. Botany 17*, 425–685.

Slatkin, M., and Futuyma, D. J. (1983). *Coevolution*. Sinauer, Sunderland, Mass.

Skutch, A. F. (1971). *A Naturalist in Costa Rica*. Univ. Florida Press, Gainesville.

Sprengel, C. K. (1793). *Das entdeckte Geheimnis der Natur im Bau und in der Befruchtung der Blumen*. Fr. Vieweg, Berlin.

Takhtajan, A. (1969). *Flowering Plants. Origin and Dispersal*. Smithsonian Inst. Press, Washington, D.C.

Tippo, O., and Stern, W. L. (1977). *Humanistic Botany*. Norton, New York.

Werth, E. (1956). *Bau und Leben der Blumen*. Enke, Stuttgart.

Willson, M. F. (1983). *Plant Reproductive Ecology*. John Wiley, New York.

CHAPTER ONE

Banks, H. P. (1970). 'Major evolutionary events and the geological record of plants: a summary'. *Biol. Rev. 47*, 451–4.

Bino, R. J., and Dafni, A. (1983). 'Entomophily and nectar secretion in the dioecious gymnosperm *Ephedra aphylla* Forssk', pp. 99–104 in: D. L. Mulcahy and E. Ottaviano, (Eds.), *Pollen: Biology and Implications for Plant Breeding*. Elsevier, New York.

— and Meeuse, A. D. J. (1981). 'Entomophily in dioecious species of *Ephedra*: A preliminary report', *Acta Bot. Neerl. 30*, 151–3.

Bold, H. C., and Wynne, M. J. (1978). *Introduction to the Algae*. Prentice-Hall, Englewood Cliffs, N.J.

Bower, F. O. (1923, 1926, 1928). *The Ferns*, vols. 1–3. Cambridge Univ. Press, London.

Crépet, W. L. (1979). 'Insect pollination: a palaeontological perspective', *Bio-Sci. 29*, 102–8.

Cronquist, A. (1981). *An Integrated System of Classification of Flowering Plants*. Columbia Univ. Press, New York.

Dilcher, D. L. (1979). 'Early angiosperm reproduction: an introductory report', *Rev. Palaeobot. Palynol. 27*, 291–328.

Ende, H. van den (1976). *Sexual Interactions in Plants*. Academic Press, New York.

Friis, E. M., and Skarby, A. (1982). '*Scandianthus*, gen. nov., Angiosperm flowers of saxifragalean affinity from the Upper Cretaceous of southern Sweden', *Ann. Bot. (NS) 50*, 569–84.

Fritsch, F. E. (1935). *The Structure and Reproduction of the Algae*. Cambridge Univ. Press, Cambridge.

Godley, E. J., and Smith, D. H. (1981). 'Breeding systems in New Zealand plants. 5. *Pseudowintera colorata* (Winteraceae)', *New Zeal. J. Bot. 19*, 151–6.

Gottsberger, G. (1974). 'The structure and function of the primitive angiosperm flower – a discussion', *Acta Bot. Neerl. 23*, 461–71.

Hickey, L. J., and Doyle, J. A. (1977). 'Early Cretaceous fossil evidence for angiosperm evolution', *Bot. Rev. 43*, 1–104.

Hughes, N. F. (1976). *Palaeobiology of Angiosperms – Problems of Mesozoic Seed-plant Evolution*. Cambridge Univ. Press, Cambridge.

Krassilov, V. A. (1977). 'The origin of angiosperms', *Bot. Rev. 43*, 143–76.

Leroi-Gourhan, A. (1975). 'The flowers found with Shanidar IV, a Neanderthal burial in Iraq', *Science 190*, 562–4.

Levine, R. P., and Ebersold, W. T. (1960). 'The Genetics and Cytology of *Chlamydomonas*', *Ann. Rev. Microbiol. 14*, 197–216.

Maynard Smith, J. (1978). *The Evolution of Sex*. Cambridge Univ. Press, Cambridge.

Meeuse, A. D. J. (1978a). 'Nectarial secretion, floral evolution, and the pollination syndrome in early angiosperms', *Proc. Kon. Ned. Akad. Wet. Amsterdam, Ser. C, 81*, 300–26.

— (1978b). 'The significance of the Gnetatae in connection with the early evolution of the angiosperms', in: *Glimpses Pl. Res.* (P. K. K. Nair, Ed.) *4*, 62–73.

— (1979). 'Why were the early Angiosperms so successful?' *Proc. Kon. Ned. Akad. Wet. Amsterdam, Ser. C, 82*, 343–69.

— (1981). 'Evolution of the Magnoliophyta: current and dissident viewpoints', pp. 393–442 in: C. P. Malik (Ed.), *Annu. Rev. Pl. Sci.*, vol. II (1980); Kalyani Publi.

Mulcahy, D. L. (1979). 'The rise of angiosperms: a genecological factor', *Science 206*, 20–3.

— (1981). 'Rise of the Angiosperms', *Nat. History 90*, 30–5.

Regal, P. J. (1977). 'Ecology and evolution of flowering plant dominance', *Science 196*, 622–9.

Retallack, G., and Dilcher, D. L. (1981). 'A coastal hypothesis for the dispersal and rise to dominance of flowering plants', pp. 27–77 in: K. Niklas (Ed.), *Palaeobotany, Palaeoecology and Evolution*, vol. 2. Praeger, New York.

Rothwell, G. W. (1977). 'Evidence for a pollination-drop mechanism in Palaeozoic pteridosperms', *Science 198*, 1251–2.

Scagel, R. F., Bandoni, R. J., Rouse, G. E., Schofield, W. B., Stein, J. R., and Taylor, T. M. C. (1966). *An Evolutionary Survey of the Plant Kingdom*, 3rd printing. Wadsworth, Belmont, Calif.

Takhtajan, A. (1969). *Flowering Plants: Origin and Dispersal*. Oliver and Boyd, Edinburgh.

Taylor, T. N. (1981). *Palaeobotany: An Introduction to Floral Plant Biology*. McGraw-Hill, New York.

Thien, L. B. (1980). 'Patterns of pollination in primitive angiosperms', *Biotropica 12*, 1–13.

Whitehead, D. R. (1969). 'Wind pollination in the angiosperms: evolutionary and environmental considerations', *Evolution 23*, 28–35.

Williams, G. C. (1975). *Sex and Evolution*. Princeton Univ. Press, Princeton, N.J.

Wolken, J. (1967). *Euglena*, 2nd edn. Appleton Century Crofts, New York.

CHAPTER TWO

Armbruster, W. S., and Webster, G. L. (1979). 'Pollination of two species of *Dalechampia* (Euphorbiaceae) in Mexico by euglossine bees', *Biotropica 11*, 278–83.

Baker, H. G., and Baker, I. (1973). 'Amino acids in nectar and their evolutionary significance', *Nature 241*, 543–5.

— and — (1979). 'Starch in angiosperm pollen grains and its evolutionary significance', *Am. J. Bot. 66*, 591–600.

— and — (1983). 'Floral nectar sugar constituents in relation to pollinator type', pp. 117–41 in: C. E. Jones and R. J. Little (Eds.). *Handb. Exp. Pollination Biol.* Van Nostrand Reinhold, New York.

Beach, J. H. (1982). 'Beetle pollination of *Cyclanthus bipartitus* (Cyclanthaceae)', *Am. J. Bot. 69*, 1074–81.

Bentley, B., and Elias, T. (1983). *The Biology of Nectaries*. Columbia Univ. Press, New York.

Chen, J., and Meeuse, B. J. D. (1971). 'Production of free indole by some Aroids', *Acta Bot. Neerl. 20*, 627–35.

Daumer, K. (1958). 'Blumenfarben, wie sie die Bienen sehen', *Z. Vergl. Physiol. 41*, 49–110.

De Vries, P. J. (1979). 'Pollen-feeding rainforest *Parides* and *Battus* butterflies in Costa Rica', *Biotropica 11*, 237–8.

Dodson, C. H., Dressler, R. L., Hills, H. G., Adams, R. M., and Williams, N. H. (1969). 'Biologically active compounds in orchid fragrances', *Science 164*, 1243–9.

Eisner, T., Silberglied, R. E., Aneshansley, D., Carrel, J. E., and Howland, H. C. (1969). 'Ultraviolet video-viewing: the television-camera as an insect eye', *Science 166*, 1172–4.

Frisch, K. von (1914). 'Der Farbensinn und Formensinn der Biene', *Zool. Jahrb. Abt. Allg. Zool. Physiol. Tiere 35*, 1–188.

Gilbert, L. E. (1972). 'Pollen feeding and reproductive biology of *Heliconius* butterflies', *Proc. Nat'l Acad. Sci. 69*, 1403–7.

Goldsmith, T. H. (1980). 'Hummingbirds see near ultraviolet light', *Science 207*, 786–8.

Gori, D. F. (1983). 'Post-pollination phenomena and adaptive floral changes', pp. 31–49 in: C. E. Jones and R. J. Little (Eds.), *Handb. Exp. Pollination Biol.* Van Nostrand Reinhold, New York.

Grant, K. A. (1966). 'A hypothesis concerning the prevalence of red coloration in California hummingbird flowers', *Am. Nat. 100*, 85–98.

— and Grant, V. (1968). *Hummingbirds and Their Flowers*. Columbia Univ. Press, New York.

Grant, V. (1950). 'The pollination of *Calycanthus occidentalis*', *Am. J. Bot. 37*, 294–7.

Ilse, D. (1928). 'Über den Farbensinn der Tagfalter', *Z. Vergl. Physiol. 8*, 658–91.

Kay, Q. O. N., Daoud, H. S., and Stirton, C. H. (1981). 'Pigment distribution, light-reflection and cell structure in petals', *Bot. J. Linn. Soc. 83*, 57–84.

Kevan, P. G. (1978). 'Floral coloration, its colorimetric analysis and significance in anthecology', pp. 51–78 in: A. J. Richards (Ed.), *The Pollination of Flowers by Insects*. Linnean Soc. Symp., No. 6 Acad. Press, London.

— (1983). 'Floral colors through the insect eye: what they are and what they mean', pp. 3–30 in: C. E. Jones and R. J. Little (Eds.), *Handb. Exp. Pollination Biol.* Van Nostrand Reinhold, New York.

Knoll, F. (1921). 'Insekten und Blumen, II. *Bombylius fuliginoaus* und die Farbe der Blumen', *Abh. Zool.-Bot. Ges. Wien 12*, 17–119.

— (1922). 'III. Lichtsinn und Blumenbesuch des Falters von *Macroglossa stellatarum*', *Abh. Zool.-Bot. Ges. Wien 12*, 120–377.

Kugler, H. (1936). 'Die Ausnützung der Saftmalsumfärbung bei den Roszkastanienblüten durch Bienen und Hummeln', *Ber. Dtsch. Bot. Ges. 60*, 128–34.

Kullenberg, B., and Bergström, G. (1973). 'Chemical communication between living organisms', *Endeavour 34*, 59–66.

Manning, A. (1956). 'The effect of honey guides', *Behaviour 9*, 114–39.

Menzel, R., and Erber, J. (1978). 'Learning and memory in bees', *Sci. Am. 239*, 102–8, 110.

Percival, M. S. (1961). 'Types of nectar in angiosperms', *New Phytol. 60*, 235–81.

Proctor, J., and Proctor, S. (1978). *Nature's Use of Color in Plants and Their Flowers*. Peter Lowe, London.

Rickson, F. R. (1979). 'Ultrastructural development of the beetle food tissue of *Calycanthus* flowers', *Am. J. Bot. 66*, 80–6.

Seogin, R. (1983). 'Visible floral pigments and pollinators', pp. 160–72 in: C. E. Jones and R. J. Little (Eds.), *Handb. Exp. Pollination Biol.* Van Nostrand Reinhold, New York.

Silberglied, R. E. (1979). 'Communication in the ultraviolet', *Annu. Rev. Ecol. Syst. 10*, 373–98.

Simpson, B. R., and Neff, J. L. (1983). 'Evolution and diversity of floral rewards', pp. 142–59 in: C. E. Jones and R. J. Little (Eds.), *Handb. Exp. Pollination Biol.* Van Nostrand Reinhold, New York.

Stanley, R. G., and Linskens, H.-F. (1974). *Pollen. Biology. Biochemistry. Management*. Springer, New York.

Thompson, W. R., Mein-Wald, J., Aneshansley, D., and Eisner, T. (1972). 'Flavonols: pigments responsible for ultraviolet absorption in nectar guides of flowers', *Science 177*, 528–30.

Timberlake, C. F., and Bridle, P. (1975). 'The anthocyanins', pp. 214–66 in: J. B. Harborne, T. J. Mabry, and H. Mabry (Eds.), *The Flavonoids*. Academic Press, New York.

Vogel, S. (1963). 'Duftdrüsen im Dienste der Bestäubung: über Bau und Funktion der Osmophoren', *Akad. Wiss. Lit. (Mainz), Abh. Math.-Naturwiss. Kl.*, Jahrgang 1962: 599–763.

143

— (1974). *Ölblumen und ölsammelnde Bienen.* Steiner, Wiesbaden.

Waddington, K. D., Allen, T., and Heinrich, B. (1981). 'Floral preferences of bumblebees (*Bombus edwardsii*) in relation to intermittent versus continuous rewards', *Animal Behav. 29*, 779–84.

Williams, N. H. (1983). 'Floral fragrances as cues in animal behavior', pp. 50–72 in: C. E. Jones and R. J. Little (Eds.), *Handb. Exp. Pollination Biol.* Van Nostrand Reinhold, New York.

CHAPTER THREE

Bawa, K. S. (1980). 'Evolution of dioecy in flowering plants', *ARES 11*, 15–39.

— and Beach, J. H. (1981). 'Evolution of sexual systems in flowering plants', *Ann. Missouri Bot. Gard. 68*, 254–74.

Beach, J. H. (1981). 'Pollinator foraging and the evolution of dioecy', *Am. Nat. 118*, 572–7.

Bertin, R. J. (1982). 'The ecology and maintenance of andromonoecy', *Evol. Theory 6*, 25–32.

Bierzychudek, P. (1982). 'Jack and Jill in the pulpit', *Nat. Hist. 91*: 22–27.

Charlesworth, B., and Charlesworth, D. (1978). 'A model for the evolution of dioecy and gynodioecy', *Am. Nat. 112*, 975–97.

— and — (1979). 'The maintenance and breakdown of distyly', *Am. Nat. 114*, 499–513.

Charlesworth D., and Charlesworth, B. (1979). 'A model for the evolution of distyly', *Am. Nat. 114*, 467–98.

Charnov, E. L., and Bull, J. J. (1977). 'When is sex environmentally determined?', *Nature 266*, 828–30.

—, Maynard-Smith, J., and Bull, J. J. (1976). 'Why be an hermaphrodite?', *Nature 263*, 125–6.

Cox, P. A. (1981). 'Niche partitioning between sexes of dioecious plants', *Am. Nat. 117*, 295–307.

Darwin, C. (1884). *The Different Forms of Flowers on Plants of the Same Species.* Reprint of 2nd edn. 1880, Appleton, New York.

Frankel, R., and Galun, E. (1977). *Pollination Mechanisms, Reproduction, and Plant Breeding.* Springer, New York.

Freeman, D. C., Harper, K. T., and Ostler, W. K. (1980). 'Ecology of plant dioecy in the intermountain region of western North America and California', *Oecologia 44*, 410–17.

—, Klikoff, L. G., and Harper, K. T. (1976). 'Differential resource utilization by the sexes of dioecious plants', *Science 193*, 597–9.

—, McArthur, E. D., Harper, K. T., and Blauer, A. C. (1981). 'Influence of environment on the floral sex ratio of monoecious plants', *Evolution 35*, 194–7.

Ganders, F. R. (1979). 'The biology of heterostyly', *New Zeal. J. Bot. 17*, 607–35.

Heslop-Harrison, J. (1975). 'Incompatibility and the pollen-stigma interaction', *Ann. Rev. Pl. Physiol. 26*, 403–25.

Horovitz, A., and Beiles, A. (1980). 'Gynodioecy as a possible populational strategy for increasing reproductive output', *Theor. Appl. Genet. 57*, 11–15.

Janzen, D. H. (1977). 'A note on optimal mate selection by plants', *Am. Nat. 111*, 365–71.

Jong, G. de (1980). 'Some numerical aspects of sexuality', *Am. Nat. 116*, 712–18.

Lewis, D. (1942). 'The evolution of sex in flowering plants', *Biol. Rev. 17*, 46–67.

Lloyd, D. G. (1979a). 'Evolution toward dioecy in heterostylous populations', *Plant. Syst. Evol 131*, 71–80.

— (1979b). 'Some reproductive factors affecting the selection of self-fertilization in plants', *Am. Nat. 113*, 67–79.

— (1980). 'Benefits and handicaps of sexual reproduction', *Evol. Biol. 13*, 69–111.

— (1982). 'Selection of combined versus separate sexes in seed plants', *Am. Nat. 120*, 571–85.

Lovett Doust, J., and Cavers, P. B. (1982). 'Sex and gender dynamics in jack-in-the-pulpit, *Arisaema triphyllum* (Araceae)', *Ecology 63*, 707–808.

Nettancourt, D. de (1977). *Incompatibility in Angiosperms.* Springer, New York.

Policansky, D. (1981). 'Sex choice and size advantage model in jack-in-the-pulpit', *PNAS(USA) 78*, 1306–8.

Ross, M. D. (1982). 'Five evolutionary pathways to subdioecy', *Am. Nat. 119*, 297–318.

Thomson, J. D., and Barrett, S. C. M. (1981). 'Selection for outcrossing, sexual selection, and the evolution of dioecy in plants', *Am. Nat. 118*, 443–9.

Vuilleumier, B. S. (1967). 'The origin and evolutionary development of heterostyly in the angiosperms', *Evolution 21*, 210–26.

Waller, D. M. (1982). 'Jewelweed's sexual skills', *Nat. Hist. 91*, 32–9.

Willson, M. F. (1982). 'Sexual selection and dicliny in Angiosperms', *Am. Nat. 119*, 579–83.

CHAPTER FOUR

Aker, C. L., and Udovic, D. (1981). 'Oviposition and pollination behavior of the yucca moth, *Tegeticula maculata* (Lepidoptera, Prodoxidae), and its relation to the reproductive biology of *Yucca whipplei* (Agavaceae), *Oecologia 49*, 96–101.

Alford, D. V. (1975). *Bumblebees.* Davis-Poynter, London.

Armstrong, J. A. (1979). 'Biotic pollination mechanisms in the Australian flora – a review', *New Zeal. J. Bot. 17*, 467–508.

Carpenter, F. L. (1983). 'Pollination energetics in avian communities: simple concepts and complex realities', pp. 215–34 in: C. E. Jones and R. J. Little (Eds.), *Handb. Exp. Pollination Biol.* Van Nostrand Reinhold, New York.

Coe, M. J., and Isaac, F. M. (1965). 'Pollination of the baobab (*Adansonia digitata* L.) by the lesser bush baby (*Galago crassicaudatus* E. Geoffroy)', *East Africa Wildlife J. 3*, 123–4.

Darwin, C. (1890). *The various contrivances by which orchids are fertilised by insects.* 2nd edn. London.

Dodson, C. H. (1975). 'Coevolution of orchids and bees' in: L. E. Gilbert and P. H. Raven (Eds.), *Coevolution of animals and Plants.* Univ. Texas Press, Austin.

Dormer, K. J. (1960). 'The truth about pollination in *Arum*', *New Phytol. 59*, 209–81.

Dressler, R. L. (1981). *The Orchids.* Harvard Univ. Press, Cambridge, Mass.

Fleming, T. H. (1982). 'The foraging strategies of plant-visiting bats', pp. 287–325 in: T. H. Kunz (Ed.), *Ecology of Bats.* Plenum, New York.

Ford, H. A., Paton, D. C., and Forde, N. (1979). 'Birds as pollinators of Australian plants', *New Zeal. J. Bot. 17*, 509–19.

Frisch, K. von (1967). *The Dance Language and Orientation of Bees.* Harvard Univ. Press. Cambridge, Mass.

Galil, J. (1977). 'Fig biology', *Endeavour 1*, 52–6.

Gilbert, L. E., and Raven, P. (1975). *Coevolution of animals and plants.* Univ. Texas Press, Austin, Texas.

Greenewalt, C. H. (1960). *Hummingbirds.* Doubleday, Garden City, N.Y.

Hainsworth, F. R. (1981). 'Energy regulation in hummingbirds', *Am. Scient. 69*, 420–9.

Heinrich, B. (1971). 'Temperature regulation in the sphinx moth, *Manduca sexta*', *J. Exp. Biol. 54*, 141–52.

— (1975). 'Thermoregulation in bumblebees. II. Energetics of warm-up and free flight', *Journal of Comparative Physiology 96*, 155–66.

— (1979). *Bumblebee Economics*. Harvard Univ. Press, Cambridge, Mass.

Heithaus, E. R. (1982). 'Coevolution between bats and plants', pp. 327–67 in: T. H. Kunz (Ed.), *Ecology of Bats*. Plenum, New York.

Hickman, J. C. (1974). 'Pollination by ants: a low-energy system', *Science 184*, 1290–2.

Howell, D. J. (1979). 'Flock-foraging in nectar-feeding bats: advantages to the bats and to the host plants', *Am. Nat. 114*, 23–49.

— (1974). 'Bats and pollen: physiological aspects of the syndrome of chiropterophily', *Comp. Biochem. Physiol. 48A*, 263–76.

— (1977). 'Time sharing and body partitioning in bat-plant pollination systems', *Nature 270*, 509–10.

Janson, C. A., Terborgh, J., and Emmons, L. H. (1981). 'Non-flying mammals as pollinating agents in the Amazonian forest', *Biotropica 13* (2), *Suppl.* 1–6.

Johnsgard, P. (1983). *The Hummingbirds of North America*. Smithsonian Inst. Press, Washington, D.C.

Kincaid, T. (1963). 'The ant plant, *Orthocarpus pusillus*', *Trans. Am. Micr. Soc. 82*, 101–5.

Knoch, E. (1899). 'Untersuchungen über die Morphologie, Biologie und Physiologie der Blüte von *Victoria regia*', *Biol. Bot. 47*, 1–60.

Kullenberg, B., and Bergström, G. (1976). 'The pollination of *Ophrys* orchids', *Bot-Notiser 129*, 11–20.

Lumer, C. (1980). 'Rodent pollination of *Blakea* (Melastomataceae) in a Costa Rican cloud forest', *Brittonia 32*, 512–17.

Marshall, A. G. (1983). 'Bats, flowers and fruit: evolutionary relationships in the Old World', *Biol. J. Linnean Soc. 20*, 115–35.

Meeuse, B. J. D. (1959). 'Beetles as pollinators', *Biologist 42*, 22–32.

— (1966). 'The voodoo lily', *Scient. Am. 215*, 80–8.

Mesler, M. R., Ackerman, J. D., and Lu, K. L. (1980). 'The effectiveness of fungus gnats as pollinators', *Am. J. Bot. 67*, 564–7.

Pijl, L. van der (1956). 'Remarks on pollination by bats in the genera *Freycinetia*, *Duabanga* and *Haplophragma*, and on chiropterophily in general', *Acta Bot. Neerl. 5*, 135–44.

— (1961). 'Ecological aspects of flower evolution. II. Zoophilous flower classes', *Evolution 15*, 44–59.

— and Dodson, C. H. (1966). *Orchid Flowers: Their Pollination and Evolution*. Univ. of Miami Press, Coral Gables, Fla.

Prime, C. T. (1960). *Lords and ladies*. Collins, London.

Seeley, T. D. (1983). 'The ecology of temperate and tropical honeybee societies', *Am. Scient. 71*, 264–72.

Stiles, F. G. (1978). 'Ecological and evolutionary implications of bird pollination', *Am. Zool. 18*, 715–27.

Stoutamire, W. P. (1968). 'Mosquito pollination of *Habenaria obtusata* (Orchidaceae)', *Michigan Bot. 7*, 203–12.

Sussman, R. W., and Raven, P. H. (1978). 'Pollination by lemurs and marsupials: an archaic coevolutionary system', *Science 200*, 731–6.

Turner, V. (1982). 'Marsupials as pollinators in Australia', in: J. A. Armstrong, J. M. Powell, and A. J. Richards (Eds.), *Pollination and Evolution*. Royal Botanic Garden, Sydney.

Vogel, S. (1966). 'Parfümsammelnde Bienen als Bestäuber von Orchideen und *Gloxinia*', *Österr. Bot. Z. 113*, 302–61.

— (1968, 1969). 'Chiropterophilie in der neotropischen Flora, I–III', *Flora 157*, 562–602; *158*, 185–222 and 269–323.

Wiens, D., and Rourke, J. P. (1978). 'Rodent pollination in southern Africa *Protea* spp.', *Nature 276*, 71–3.

Williams, N. H. (1982). 'The biology of orchids and euglossine bees', pp. 119–71 in: J. Arditti (Ed.). *Orchid Biology, II*. Cornell Univ. Press, Ithaca, N.Y.

CHAPTER FIVE

Ackerman, J. D. (1981). 'Pollination biology of *Calypso bulbosa* var. *occidentalis* (Orchidaceae): a food-deception system', *Madroño 28*, 101–10.

Bawa, K. S. (1980). 'Mimicry of male by female flowers and intrasexual competition for pollinators in *Jacaratia dolichaula* (D. Smith) Woodson (Caricaceae)', *Evolution 34*, 467–74.

Bierzychudek, P. (1981). '*Asclepias*, *Lantana*, and *Epidendrum*: a floral mimicry complex?' *Reprod. Biology*, 54–8.

Bobisud, L. E., and Neuhaus, R. J. (1975). 'Pollinator constancy and the survival of rare species', *Oecologia 21*, 263–72.

Boyden, T. C. (1980). 'Floral mimicry by *Epidendrum ibaguense* (Orchidaceae) in Panama', *Evolution 34*, 135–6.

Brown, J. H., and Kodric-Brown, A. (1979). 'Convergence, competition, and mimicry in a temperate community of hummingbird-pollinated flowers', *Ecology 60*, 1022–35.

Coleman, E. (1928). 'Pollination of an Australian orchid by the male ichneumonid *Lissopimpla semipunctata* Kirby', *Trans. Ent. Soc. Lond. 76*, 533–9.

Colwell, R. K., Betts, B. J., Bunnell, P., Carpenter, F. L., and Feinsinger, P. (1974). 'Competition for the nectar of *Centropogon Valerii* by the hummingbird *Colibri thalassinus* and the flower-piercer *Diglossa plumbea*, and its evolutionary implications', *Condor 76*, 447–52.

Correvon, H., and Pouyanne, M. (1916). 'Un curieux cas de mimétisme chez les Ophrydées', *J. Soc. Nat. Hortic. France 29*, 23–84.

Dafni, A., and Ivri, Y. (1981a). 'Floral mimicry between *Orchis israelitica* Baumann and Safri (Orchidaceae) and *Bellevalia flexuosa* Boiss. (Liliaceae)', *Oecologia 49*, 229–32.

— and — (1981b). 'The flower biology of *Cephalanthera longifolia* (Orchidaceae): pollen imitation and facultative floral mimicry', *Pl. Syst. Evol. 137*, 229–40.

Fritz, R. S., and Morse, D. H. (1981). 'Nectar parasitism of *Asclepias syriaca* by ants: Effect on nectar levels, pollinia insertion, pollinaria removal and pod production', *Oecologia 50*, 316–19.

Guerrant, E. O., and Fiedler, P. L. (1981). 'Flower defenses against nectar pilferage by ants', *Biotropica 13* (2), *Suppl.*: 25–33.

Inouye, D. W. (1980). 'The terminology of floral larceny', *Ecology 61*, 1251–3.

Kerner, A. (1878). *Flowers and Their Unbidden Guests* (trans. and ed. by W. Ogle). Paul, London.

Kullenberg, B. (1961). *Studies in Ophrys Pollination*. Almqvist and Wiksells, Uppsala.

— and Bergström, G. (1976). 'The pollination of *Ophrys* orchids', *Bot. Notiser 129*, 11–19.

Little, R. J. (1980). 'Floral mimicry between two desert annuals, *Mohavea confertiflora* (Scrophulariaceae) and *Mentzelia involucrata* (Loasaceae)', Ph.D. thesis, Claremont, Calif.

— (1983). 'A review of floral food deception mimicries with comments on floral mutualism', pp. 249–309 in: C. E. Jones and R. J. Little (Eds.), *Handb. Exp. Pollination Biol.* Van Nostrand Reinhold, New York.

McDade, L. A., and Kinsman, S. (1980). 'The impact of floral parasitism in two neotropical hummingbird-pollinated plant species', *Evolution 34*, 944–58.

Meeuse, B. J. D., and Schneider, E. L. (1980). '*Nymphaea* revisited: a preliminary communication', *Israel J. Bot. 28*, 65–79.

Powell, E. A., and Jones, C. E. (1983). 'Floral mutualism in *Lupinus benthamii* (Fabaceae) and *Delphinium parryi* (Ranunculaceae)', pp. 310–29 in: C. E. Jones and R. J. Little (Eds.), *Handb. Exp. Pollination Biol.* Van Nostrand Reinhold, New York.

Roubik, D. W. (1982). 'The ecological impact of nectar-robbing bees and pollinating hummingbirds on a tropical shrub', *Ecology 63*, 354–60.

Rust, R. W. (1979). 'Pollination of *Impatiens capensis*: Pollinators and nectar robbers', *J. Kansas Entomol. Soc. 52*, 297–308.

Schemske, D. W. (1981). 'Floral convergence and pollinator sharing in two bee-pollinated tropical herbs', *Ecology 62*, 946–54.

Schremmer, F. (1941). 'Eine Bauchsammelbiene (*Megachile circumcincta*) als Zerstörerin der Blüten von *Salvia glutinosa*', *Zool. Anz. 133*, 230–2.

— (1955). 'Über anormalen Blütenbesuch und das Lernvermögen blütenbesuchender Insekten', *Österr. Bot. Z. 102*, 551–71.

Stephenson, A. G. (1981). 'Toxic nectar deters nectar thieves of *Catalpa speciosa*', *Am. Midl. Nat. 105*, 381–3.

Stoutamire, W. P. (1974). 'Australian terrestrial orchids, thynnid wasps, and pseudocopulation', *Am. Orchid Soc. Bull. 1974*, 13–18.

Vane-Wright, R. J. (1980). 'On the definition of mimicry', *Biol. J. Linnean Soc. 13*, 1–6.

Wickler, W. (1968). *Mimicry in Plants and Animals* (World Univ. Library). McGraw-Hill, New York.

Wiens, D. (1978). 'Mimicry in plants', in: M. K. Hecht, W. C. Steere, and B. Wallace (Eds.), *Evolutionary Biology*. vol. 11, Plenum, New York.

Williamson, G. B., and Black, E. M. (1981). 'Mimicry in hummingbird-pollinated plants?', *Ecology 62*, 494–6.

Wyatt, R. (1980). 'The impact of nectar-robbing ants in the pollination system of *Asclepias curassavica*', *Bull. Torrey Bot. Club 107*, 24–8.

CHAPTER SIX

Baker, H. G. (1955). 'Self-compatibility and establishment after "long-distance" dispersal', *Evolution 9*, 347–8.

Cruden, R. W., and Miller-Ward, S. (1981). 'Pollen-ovule ratio, pollen size, and the ratio of stigmatic area to the pollen-bearing areas of the pollinator: An hypothesis', *Evolution 35*, 964–74.

Darwin, C. (1876). *The effects of cross- and self-fertilisation in the vegetable kingdom.* London.

Daumann, E. (1963). 'Zur Frage nach dem Ursprung der Hydrogamie, zugleich ein Beitrag zur Blütenökologie von *Potamogeton*', *Preslia 35*, 23–30.

— (1966). 'Pollenkitt, Bestäubungsart und Phylogenie', *Novit. Bot. Univ. Carol. Pragensis.* 1966, 19–28.

— (1970). 'Zur Frage nach der Bestäubung durch Regen (Ombrogamie)', *Preslia 42*, 220–4.

Dyakowska, J., and Zarzycki, J. (1959). 'Gravimetric studies on pollen', *Bull. Acad. Pol. Sci. Cl. II, Sér. Sci. Biol. 7*, 11–16.

Ernst-Schwarzenbach, M. (1945). 'Zur Blütenbiologie einiger Hydrocharitaceen', *Ber. Schweiz. Bot. Ges. 55*, 33–69.

Faegri, K., and Iversen, J. (1975). *Textbook of pollen analysis*, 3rd edn. Munksgaard, Copenhagen.

Fryxell, P. G. (1957). 'Mode of reproduction of higher plants', *Bot. Rev. 23*, 135–233.

Hagerup, O. (1954). 'Autogamy in some drooping Bicornes flowers', *Bot. tidskr. 51*, 103–6.

Hartog, C. den (1964). 'Over de oecologie van bloeiende *Lemna trisulca*', *Gorteria 2*, 68–72.

Haumann-Merck, L. (1912). 'Observations éthologiques et systématiques sur deux espèces argentines du genre *Elodea*', *Rec. Inst. Bot. Leo Errera 9*, 33–9.

Jong, P. C. de (1976). '*Flowering and Sex-expression in Acer*', *Meded. Landb. Hogeschool Wageningen 76*, 2.

Kaplan, S. M., and Mulcahy, D. L. (1971). 'Mode of pollination and floral sexuality in *Thalictrum*', *Evolution 25*, 659–68.

Knoll, F. (1930). 'Über Pollenkitt und Bestäubungsart', *Z. Bot. 23*, 609–75.

Koski, V. (1970). 'A study of pollen dispersal as a mechanism of gene flow in conifers', *Commun. Inst. Forest Fenn. 70*, 4.

Kugler, H. (1975). 'Die Verbreitung anemogamer Arten in Europa', *Ber. Dtsch. Bot. Ges. 88*, 441–50.

Mahabale, T. S. (1968). 'Spores and pollen of water plants and their dispersal', *Rev. Palaeobot. Palynol. 7*, 285–96.

Mosebach, G. (1932). 'Über die Schleuderbewegungen der explodierenden Staubgefässe und Staminodien bei einigen Urticaceen', *Planta 16*, 70–115.

Pande, G. K., Pakrash, R., and Hassam, M. A. (1972). 'Floral biology of barley (*Hordeum vulgare* L.)', *Ind. J. Agric. Sci. 48*, 697–703.

Pettitt, J., Ducker, B., and Knox, B. (1981). 'Submarine pollination', *Scient. Am. 244*, 134–43.

Pojar, J. (1973). 'Pollination of typically anemophilous salt marsh plants by bumblebees, *Bombus terricola occidentalis* Gren.', *Am. Midl. Nat. 89*, 448–51.

Rempe, H. (1937). 'Untersuchungen über die Verbreitung des Blütenstaubes durch die Luftströmungen', *Planta 27*, 93–147.

Schemske, D. W. (1978). 'Evolution of reproductive characteristics in *Impatiens* (Balsaminaceae): The significance of cleistogamy and chasmogamy', *Ecology 59*, 596–613.

Stebbins, G. L. (1957). 'Self-fertilization and population variability in higher plants', *Am. Nat. 91*, 337–54.

Uphof, J. C. T. (1938). 'Cleistogamic Flowers', *Bot. Rev. 4*, 21–50.

Wells, H. (1979). 'Self-fertilization: Advantageous or deleterious?', *Evolution 33*, 252–5.

Wodehouse, R. P. (1935). *Pollen Grains.* McGraw-Hill, New York.

— (1945). *Hayfever Plants.* Chronica Botanica, Waltham, Mass.

CHAPTER SEVEN

Allard, R. W. (1960). *Principles of Plant Breeding.* John Wiley & Sons, New York.

Arditti, J. (1977). 'Clonal propagation of orchids by means of tissue culture – A manual', pp. 203–93 in: J. Arditti (Ed.), *Orchid Biology, I.* Cornell Univ. Press, Ithaca, N.Y.

—, Clements, M. A., Fast, G., Hadley, G., Nishimura, G., and Ernst, R. (1982). 'Orchid seed germination and seedling culture – A manual', pp. 243–370 in: J. Arditti (Ed.), *Orchid Biology, II.* Cornell Univ. Press, Ithaca, N.Y.

Atkins, E. L., Jr., Anderson, L. D., and Greywood, E. A. (1970). 'Research on the effect of pesticides on honeybees 1968–69', *Am. Bee J. 110*, 387–9.

Baker, H. G. (1978). *Plants and Civilization*, 3rd edn. Wadsworth, Belmont, Cal.

Barnes, D. K. (1980). 'Alfalfa', pp. 177–87 in: *Hybridization of Crop Plants*. American Society of Agronomy and Crop Science, Madison, Wisc.

Beard, B. H. (1981). 'The sunflower crop', *Sci. Am.*, May, 150–61.

Bekey, R., and Klostermeyer, E. C. (1981). 'Orchard mason bee', *Washington State Univ. Extension Bulletin 0922*, 1–4.

Bohart, G. E., and Todd, F. E. (1961). 'Pollination of seed crops by insects', pp. 240–6 in: *Seeds*. Yearbook of Agric., USDA.

Cocking, E. C. (1979). 'Parasexual reproduction in flowering plants', *New Zeal. J. Bot. 17*, 665–71.

Condit, J. J., and Swingle, W. T. (1947). *The Fig*. Chronica Botanica Co., Waltham, Mass.

Crisp, P. (1976). 'Trends in breeding and cultivation of cruciferous crops', in: J. G. Vaughan, A. J. MacLeod, and B. M. G. Jones (Eds.), *The Biology and Chemistry of the Cruciferae*. Academic Press, London.

Davis, D. D. (1978). 'Hybrid cotton: specific problems and potentials', *Adv. Agron. 30*, 129–57.

Dedio, W., and Putt, E. D. (1980). 'Sunflower', pp. 631–44 in: *Hybridization of Crop Plants*. Am. Soc. Agron. Crop Sci., Madison, Wisc.

Erickson, E. H., Jr. (1983). 'Pollination of entomophilous hybrid seed parents', pp. 493–535 in: E. C. Jones and R. J. Little (Eds.), *Handb. Exp. Pollination Biol*. Van Nostrand Reinhold, New York.

Estes, J., Amos, B. B., and Sullivan, J. R. (1983). 'Pollination from two perspectives: The agricultural and biological sciences', pp. 536–54 in: E. C. Jones and R. J. Little (Eds.), *Handb. Exp. Pollination Biol*. Van Nostrand Reinhold, New York.

Fehr, W. R., and Hadley, H. H. (Eds.) (1980). *Hybridization of Crop Plants*. Am. Soc. Agron. Crop Sci., Madison, Wisc.

Fick, G. N. (1978). 'Breeding and genetics', pp. 279–338 in: *Sunflower Science and Technology*. Am. Soc. Agron. Crop Sci., Madison, Wisc.

Frankel, R., and Galun, E. (1977). *Pollination Mechanisms, Reproduction and Plant Breeding*. Springer-Verlag, Berlin.

Free, J. B. (1970). *Insect Pollination of Crops*. Academic Press, New York.

Gary, N. E. (1975). 'Activities and behavior of honeybees', pp. 185–264 in: *The Hive and the Honeybee*. Dadant & Sons, Hamilton, Ill.

Greathead, D. J. (1983). 'The multi-million dollar weevil that pollinates oil palms', *Antenna 7* (3), 105–7.

Heiser, C. B., Jr. (1981). *Seed to Civilization*, 2nd edn. Freeman, San Francisco.

Holm, S. N. (1973). '*Osmia rufa* L. (Hymenoptera: Megachilidae) as a pollinator of plants in greenhouses', *Entomol. Scand. 4*, 217–24.

Johansen, C. A., Mayer, D. F., and Eves, J. D. (1978). 'Biology and management of the alkali bee, *Nomia melanderi* Cockerell (Hymenoptera: Halictidae)', *Melanderia 28*, 23–46.

—, Eves, J. D., Mayer, D. F., Bach, J. C., Nedrow, M. E., and Kions, C. W. (1981). 'Effects of ash from Mt. St. Helens on bees', *Melanderia 37*, 20–9.

Jones, C. E., and Little, R. J. (Eds.) (1983). *Handbook of Experimental Pollination Biology*. Van Nostrand Reinhold, New York.

Kessler, K. (1980). 'Honeybees sweeten sunflower profits', *The Furrow 85*, 14–15.

Klein, Richard M. (1979). *The Green World: An Introduction to Plants and People*. Harper & Row, New York.

Mangelsdorf, P. C. (1974). *Corn: its Origin, Evolution and Improvement*. Harvard Univ. Press, Cambridge, Mass.

Martin, E. C. (1975). 'The use of bees for crop pollination', pp. 579–614 in: *The Hive and the Honeybee*. Dadant & Sons, Hamilton, Ill.

McGregor, S. E. (1976). *Insect Pollination of Cultivated Crop Plants*. Agr. Handb. No. 496, USDA.

Nitsch, J. P., and Nitsch, C. (1969). 'Haploid plants from pollen grains', *Science 163*, 85–7.

NRCC (1981). 'Pesticide-pollinator interactions', *NRCC/CNRC Publ. No. 18471, Environ. Secr. Nat'l Res. Council Canada*, Ottawa.

Prescott, W. H. (no date). *History of the Conquest of Mexico*. The Modern Library, New York.

Rubis, D. (1970). 'Breeding insect-pollinated crops', pp. 19–24 in: *The Indispensable Pollinators*. IX Pollin. Conf., Hot Springs, Ark.

Robinson, R. W., Shannon, S., and Guardia, M. D. de la (1969). 'Regulation of sex expression in the cucumber', *Bioscience 19*, 141–2.

Stuber, C. W. (1980). 'Mating designs, field nursery layouts, and breeding records', pp. 83–104 in: Am. Sci. Agron. Crop Sci., Madison, Wisc.

Syed, R. A. (1979). 'Studies on oil palm pollination by insects', *Bull. Entomol. Res. 69*, 213–24.

Tippo, O., and Stern, W. L. (1977). *Humanistic Botany*. Norton, New York.

Torchio, P. F. (1976). 'Use of *Osmia lignaria* Say (Hymenoptera: Apoidea, Megachilidae) as a pollinator in an apple and fruit orchard', *J. Kansas Entomol. Soc. 49* (4), 475–84.

Toxopeus, H. (1969). 'Cacao', pp. 79–109 in: F. P. Ferwerda and F. Wit. (Eds.), *Outlines of Perennial Crop Breeding in the Tropics*. H. Veenman & Zonen, Wageningen.

Tsunoda, S., Hinata, K., and Gomez-Campo, C., (Eds.) (1980). *Brassica Crops and Wild Allies*. Japan Sci. Soc. Press, Tokyo.

ACKNOWLEDGMENTS

When, in the year 1961, somebody asked me how long it had taken me to write my little book *The Story of Pollination*, I truthfully answered: forty-five years. Now, in 1984, the time has come to make acknowledgments concerning my present book, and one can easily see that it makes more sense than ever to mention my earliest teachers first. So, I gladly seize this opportunity to thank my parents, Adrianus Meeuse and Jannigje Kruithof, who have guided my very first steps. I also owe a tremendous debt of gratitude to those individuals who were my fellow biology students at the Universities of Leiden and Delft. They have taught me a great deal. Some, alas, are no longer with us. With deepest respect I salute the memory of Lukas Tinbergen, who has been like a brother to me. In the area of field- and pollination-biology, his older brother Niko, and that wise and warm 'elder statesman', L. van der Pijl, have probably influenced me the most.

Bastiaan Meeuse
University of Washington, Seattle

In 1973 Professor Niko Tinbergen recommended that my colleagues and I at Oxford Scientific Films should make a film about pollination, and that we should ask Professor Bastiaan Meeuse to act as consultant and scientific advisor. That simple act was the start of a journey of discovery and adventure that spanned nearly eight years and four continents. Throughout, my activities were guided by Bastiaan Meeuse's encyclopedic knowledge of all matters connected with pollination, and fuelled by his tireless enthusiasm.

It may seem strange to acknowledge a co-author so I must point out that my contribution to this book is a very minor one.

I would like to thank Niko Tinbergen for providing the spark, and Bastiaan Meeuse for sustaining the flame.

Sean Morris
Oxford Scientific Films

PICTURE CREDITS

G. I. Bernard/OSF: endpapers, 11, 14, 17, 25, 27 (*above*), 29, 33, 37, 44 (*below*), 45, 48, 51, 56, 63, 67 (*above*), 91, 99 (*above left, above right* and *below*), 116, 117, 126
Tom Boyden: 93 (*below*)
Robin Buxton/OSF: 31
M. J. Coe/OSF: 92 (*above* and *below*)
J. A. L. Cooke/OSF: 6, 35, 38 (*left*), 67 (*below*), 82, 84, 101, 106, 113, 115
Stephen Dalton/OSF: 40, 86, 108
M. P. L. Fogden/OSF: 26
Lloyd Goldwasser: 55
Dr C. E. Jeffree: 118 (*above left*)
Breck P. Kent/Animals Animals/OSF: frontispiece
Dr Kerr/Dr B. E. Juniper: 38 (*right*), 118 (*above right, below left* and *below right*)
Prof. Roger M. Knutson: 93 (*above*)
Robert W. Mitchell: 74 (*right*)
Ian Moar/OSF: 27 (*below*)
Sean Morris/OSF: 13, 32, 39, 42 (*above*), 44 (*above*), 52, 53 (*above*), 61, 66, 75, 79, 94, 96, 123
Alan G. Nelson/Earth Scenes/OSF: 132
OSF: 73
Peter Parks/OSF: 9, 99 (*centre right*)
Dr R. Parks/OSF: 18
L. Rhodes/Animals Animals/OSF: 124
David Shale/OSF: 88
Perry D. Slocum/Earth Scenes/OSF: 28
D. J. Stradling/OSF: 42 (*below*)
David Thompson/OSF: 20, 60, 74 (*left*), 104, 130, 138
Gerald Thompson/OSF: 81
Robert A. Tyrrell: 76 (*left* and *right*)
Stefan Vogel: 53 (*below*)
A. G. (Bert) Wells/OSF: 77

The artwork reference for the illustration on p. 137 was kindly provided by Eric Crichton Photos.

INDEX